Groundwater Remediation and Petroleum

A Guide for Underground Storage Tanks

David C. Noonan
James T. Curtis

 LEWIS PUBLISHERS

Library of Congress Cataloging-in-Publication Data

Noonan, David C.
   Groundwater remediation and petroleum: a guide for
underground storage tanks/by David C. Noonan and James T. Curtis.
      p.   cm.
   Includes bibliographical references.
   1. Oil pollution of water.   2. Oil storage tanks—
Environmental aspects.   3. Water, Underground—Purification.
I. Curtis, James T.   II. Title.
TD427.P4N56   1990
628.1′6833—dc20   89-48694
ISBN 0-87371-217-X

Photography by Heather Pillar.

LEWIS PUBLISHERS
121 South Main Street, P.O. Drawer 519, Chelsea, Michigan 48118

PRINTED IN THE UNITED STATES OF AMERICA

We dedicate this book to our parents.

PREFACE

This book was written to get information on cleaning up groundwater out to the many people charged with the responsibility of addressing contamination caused by leaking underground storage tanks (USTs). People faced with cleanup decisions need answers to questions: How extensive is the contamination? Which technology, or combination of technologies, should be used to remove the petroleum hydrocarbons that have reached the water table? How long will it take to clean the groundwater? How much will it cost? What design changes could be made to make the treatment technology more cost-effective? This book provides answers to these questions so decisionmakers can understand the implications—cost, design, and otherwise—of undertaking a particular course of cleanup action.

The amount of petroleum hydrocarbons that reaches the water table is often only a small fraction of the overall mass of the release from an underground storage tank. Yet this fraction is often the most troublesome and expensive to remove. One gallon of gasoline is enough to render one million gallons of groundwater unusable based on federal drinking water standards. It is through groundwater that contaminants can seep beneath homes, be drawn into wells, and enter the homes of millions of Americans. Groundwater users are at risk when groundwater becomes contaminated.

Fortunately, technologies exist to remove petroleum hydrocarbons from drinking water supplies. Packed air towers and granular activated carbon filters are widely used throughout the country to treat water contaminated by UST releases. The engineering designs, costs, and limitations associated with these technologies are widely understood. This book presents the collective experiences and knowledge gained from numerous engineering applications of these technologies. We have tried to present the information in a logical, easy-to-follow format so that readers can use this knowledge to their own benefit and apply these technologies in an efficient and cost-effective manner.

ACKNOWLEDGMENTS

The information in this book came from a variety of contributors at Camp Dresser & McKee Inc. Many of the cost data for packed air towers and carbon treatment were developed by Mr. Stephen Medlar. Mr. Tom A. Pedersen provided valuable information on biorestoration. Mr. William Glynn and Ms. Mary Tabak contributed to the development of the sample problem for the packed air tower design. Ms. Julia Nault, Mr. David Doyle, and Dr. Myron Rosenberg developed the schedule of federal UST regulations provided in Appendix A. Graphics were prepared by Mr. A. Russell Briggs and Ms. Lori Hoffer. Some of the information contained in this book was collected as part of a research assignment (Contract No. 68-01-7383) for the Environmental Protection Agency's Office of Underground Storage Tanks and is used here with permission.

 **David C. Noonan** is a Principal Engineer with Camp Dresser & McKee Inc., an international consulting engineering firm based in Boston. Mr. Noonan has been responsible for a wide variety of projects involving the cleanup of petroleum products from the subsurface. He has conducted extensive research for EPA's Office of Underground Storage Tanks, and is presently developing a groundwater remediation plan to clean up one of the largest gasoline spills in the history of New York state. He has authored numerous technical papers and guidance manuals.

Mr. Noonan received his BS in civil engineering from Northeastern University and his MBA from New Hampshire College. He serves on the editorial board of *Civil Engineering Practice*, the journal of the Boston Society of Civil Engineering, and is a registered professional engineer.

 **James T. Curtis** is a Water Resources Engineer with Camp Dresser & McKee Inc. in Boston. As a member of CDM's Water Resources Group, Mr. Curtis has been involved with several projects related to subsurface petroleum remediation for government clients, including EPA's Office of Underground Storage Tanks and the Office of Research and Development, as well as private clients. Mr. Curtis provided extensive support to the Office of Underground Storage Tanks during the development and promulgation of the federal underground storage tank regulations. Mr. Curtis received a BS in engineering from Princeton University.

CONTENTS

FIGURES

TABLES

1   INTRODUCTION

1.1 BACKGROUND

Gasoline is a complex mixture of hydrocarbons, principally comprised of alkane, alkene, and aromatic hydrocarbons. Gasoline spilled or leaked into soil volatilizes due to its high vapor pressure, filling pore spaces with its vapors. Gasoline vapors present in soil, as well as gasoline present in soil in the liquid phase, are subject to further dispersal and migration as precipitation moves into and through the subsurface. Gasoline in either of these states can become dissolved in water and eventually move to groundwater supplies.

Cleaning up a release from an underground storage tank (UST) requires both short-term emergency measures and long-term corrective actions. Short-term emergency measures involve taking immediate steps to abate imminent safety and health hazards, including potential explosions. They include notifying appropriate government agencies, stopping the release, and removing hazardous substances as necessary to prevent further releases and to allow inspection and repair of the tank system.

The focus of this book is on long-term remediation and site restoration. The task at hand is to clean up the gasoline that has been released from the tank and has dissolved in groundwater. A variety of corrective actions can be used at a leaking UST site to remove contaminants dissolved in groundwater. They differ in cost, removal efficiencies, reliability, and applicability.

There are also cross-media contamination issues associated with corrective actions that must be addressed in developing a corrective action plan for a particular site: for example, vapor emissions from air stripping towers and soil venting systems.

1.2   OBJECTIVES

The objective of this book is to provide engineering-related information regarding the costs, removal efficiencies, and limitations of alternative corrective action technologies aimed at removing contaminants (principally associated with gasoline) from groundwater. While numerous technologies exist to treat water contaminated with petroleum hydrocarbons, only a few possess demonstrated performance records and have progressed to full-scale applications. This book focuses on those corrective action technologies that (1) have been widely proven to be effective and reliable or (2) are promising technologies that appear likely to be effective but lack full-scale applications and review.

Ultimately this book can serve as a reference document or engineering manual to local and state personnel who will be required to make decisions regarding the most appropriate corrective actions to use at a particular site. The cost curves, design equations, and related implementation issues will assist consultants, engineers, and regulators in making informed and effective decisions.

The overall intent of this book is to provide personnel involved with corrective actions at USTs with a summary of the principal components, design considerations, and costs behind the technologies, and to identify conditions and situations where one corrective action might be preferred over another. *This is not a design manual, and should not be used as such.* Each situation is unique, and a professional engineer or similarly qualified individual should design and install any equipment described in this book.

To compare and contrast the various corrective action techniques, each proven technology is evaluated on the basis of several important criteria:

1. *Effectiveness*—How effective is the technology in removing contaminants?

2. *Cost*—What are the capital and operation and maintenance (O&M) costs of the technologies? What are the projected service lives? How does cost vary with time and removal efficiency?

3. *Reliability*—How consistently can the technologies remove the contaminants of concern and over how long a period of time?

4. *Ease of Operation*—Are specially trained personnel required for O&M activities? How complex is the technology?

5. *Limitations*—What factors might reduce the effectiveness or reliability of a technology, or limit its applicability in a given situation?

1.3 LIMITATIONS OF STUDY

This book focuses on widely applied and proven technologies, ones that could be recommended at a site to secure desired results. In addition, the technologies are described with a specific focus on removing gasoline from the subsurface, especially the major constituents of concern in gasoline: benzene, toluene, and xylene (BTX).

There are a number of regulatory issues associated with the implementation of corrective action technologies. These issues include securing the appropriate permits and ultimately disposing of recovered free product and groundwater. This book does not address these regulatory issues, except for those discussed under the "Limitations" section for each technology. Rather, the focus of this study is on engineering-related considerations for each technology, should one or more of these technologies be undertaken.

Other limitations on the analyses include:

- *Uncertainty about the composition of gasoline*: Over 240 compounds have been identified in gasoline. Reliable data on toxicity, chemical characteristics, and weight in gasoline do not exist for every constituent. In addition, there are a number of proprietary additives for which little or no data exist. In this study we have focused corrective actions on cleaning up the principal constituents of concern for which reliable data exist.

- *Uncertainty about site-specific conditions*: Although there is a considerable amount of literature on all of the technologies, performance and design considerations are highly dependent on site-specific conditions. To optimize system performance, field and pilot testing should be undertaken at individual sites prior to full implementation. Recognizing the need for generalizations, we have tried to base our analyses on "typical contamination incidents" and the typical concentrations one is likely to encounter in a leaking UST situation.

The book is structured as follows: Chapter 2 describes how packed air towers, carbon adsorption systems, and biorestoration techniques remove petroleum hydrocarbons from groundwater. Chapter 3 provides a detailed example problem on designing a packed air tower. Finally, Appendix A summarizes the recently enacted federal regulations governing underground storage tank systems.

2   REMOVING GASOLINE DISSOLVED IN GROUNDWATER

There are several methods of removing gasoline constituents dissolved in groundwater, including air stripping, activated carbon adsorption, biorestoration, resin adsorption, reverse osmosis, ozonation, oxidation with hydrogen peroxide, ultraviolet irradiation, and land treatment. All of these methods are capable of removing, destroying, or detoxifying all or some of the gasoline contaminants under the right circumstances. Air stripping and activated carbon adsorption, however, are the most cost-effective and widely applied in actual practice. Air stripping and/or activated carbon adsorption are applicable to most of the cases where gasoline has contaminated local groundwater. They offer the highest level of effectiveness in reducing contaminants to low levels over a wide range of situations, as well as being fairly cost-effective. Biorestoration is a technology that has only recently begun to receive attention; although promising, it has yet to be proven as a viable widespread method for controlling groundwater contaminants. Other methods such as those listed above may be effective in certain situations, but are rarely used at leaking UST sites because of their limited effectiveness and/or their high cost. In some cases, a combination of methods (e.g., air stripping followed by activated carbon adsorption) has proven to be the best and most cost-effective approach to reducing groundwater contaminants to acceptable levels. The following sections address air stripping, carbon adsorption, and biorestoration in terms of their operation, removal efficiencies, cost-effectiveness, and limitations.

2.1   AIR STRIPPING

2.1.1   Background

Air stripping is a proven, effective means to remove volatile organic chemicals from groundwater. It works by providing intimate contact of air and water to allow the diffusion of volatile substances from the liquid phase to the gaseous phase. In many cases it is the most cost-effective option for gasoline-contaminated groundwater. It has been used at many sites, either alone or in conjunction with other methods (usually activated carbon) with good results. There are several methods of air stripping. They include diffused aeration, tray aerators, spray basins, and packed towers.

Diffused Aeration

In a diffused aeration system, air (usually compressed air) is injected into the water through a diffuser or sparging device that produces fine bubbles. Mass transfer occurs across the air-water interface of the bubbles until they leave the water or become saturated with contaminant. This type of aeration is usually conducted in some type of contact chamber, although it can take place in holding ponds. A schematic diagram of a diffused aerator appears in Figure 1.

Gas transfer rates can be improved by producing smaller bubbles, increasing the air-water ratio, improving basin geometry, and using a turbine to increase turbulence. Increasing the depth of the tank will

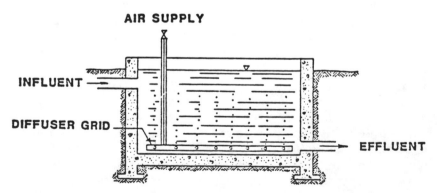

Figure 1.   Schematic of a typical diffused aerator. *Source:* Hess et al., 1983. (Reprinted from *Occurrence and Removal of Volatile Organic Chemicals from Drinking Water,* by permission. Copyright 1983, American Water Works Association Research Foundation.)

also improve the mass transfer rate if the bubbles do not reach saturation before exiting to the atmosphere (Kavanaugh and Trussell, 1981).

In practice, diffused aerators have removal efficiencies in the range of 70-90% for organics such as trichloroethylene, carbon tetrachloride, tetrachloroethylene, vinyl chloride, and others (Kavanaugh and Trussell, 1981, Dyksen et al., 1982). In some cases this may be an acceptable level of treatment; where high removal rates are required, though, diffused aerators are not practical (Figure 2). In general, this technique is less efficient and more expensive than packed towers.

Tray Aeration

Tray aeration is a simple, low-maintenance method of aeration that does not used forced air. Water is allowed to cascade through several

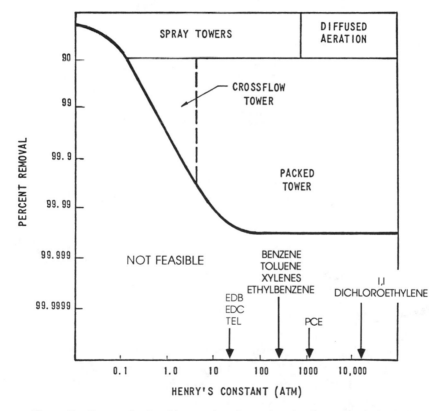

Figure 2. Ranges for feasible aeration alternatives for the removal of volatile compounds. *Source:* Hess et al., 1983. (Reprinted from *Occurrence and Removal of Volatile Organic Chemicals from Drinking Water,* by permission. Copyright 1983, American Water Works Association Research Foundation.)

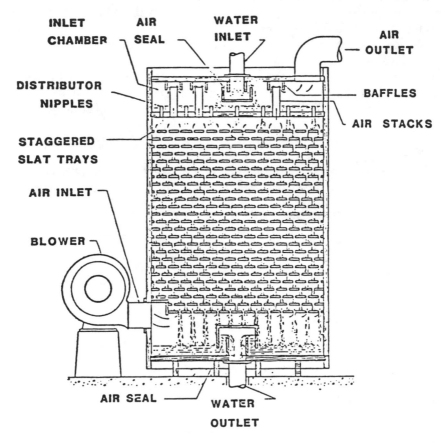

Figure 3. Schematic diagram of redwood slat tray aerator.
Source: Hess et al., 1983. (Reprinted from *Occurrence and Removal of Volatile Organic Chemicals from Drinking Water,* by permission. Copyright 1983, American Water Works Association Research Foundation.)

layers of trays or slats to increase the surface area available to the atmosphere (Figure 3). Full-scale tray aerators used for the removal of trichloroethylene, tetrachloroethylene, *trans*-1,2-dichloroethylene, 1,2-dichloroethane, and other chemicals have shown removal efficiencies of 10-90%; usual values are between 40-60% (Hess et al., 1983). In certain situations, tray aeration could be a cost-effective method of reducing VOC concentrations somewhat (for example, prior to activated carbon treatment). Like diffused aeration, this method cannot be used where low effluent concentrations are required, and has not been widely applied at leaking UST sites.

Spray Aeration

Spray aeration involves setting up a grid network of piping and nozzles over a pond or basin. Contaminated water is simply sprayed through the nozzles and into the air to form droplets. Mass transfer of the contaminant takes place across the air-water surface of the droplets. Mass transfer efficiency can be increased by passing the water through the nozzles a number of times; in Suffern, New York, removal of 1,1,1-trichloroethane was increased from 40% to 85% by passing water through the nozzles 2.5 times instead of just once (Hess et al., 1983). Spray aeration has also been used as a means of aquifer recharge; water treated by GAC and air stripping was sprayed over an 8-acre area to recharge the aquifer being cleaned (McIntyre et al., 1986). This technique could result in higher removal of VOCs (through the spraying), and an increased rate of aquifer restoration due to the recharge. Two disadvantages of spray aeration are the large land area necessary for the spray pond and the formation of large amounts of mist, which could be carried into nearby residential areas. Also, the possibility of ice formation (both from the mist and on the nozzles) would lower the usefulness of this technique in colder climates.

Packed Towers

Packed towers use a packing material to provide a high air-water ratio and a large void volume, which can result in removal efficiencies greater than those attainable by any other aeration technique. This method involves passing water down through a column of packing material while pumping air countercurrent up through the packing (Figure 4). The packing material serves to break the water into small droplets, causing a large surface area across which mass transfer can take place.

Countercurrent packed towers are the most common of the aeration methods; in fact, the term "air stripping" is often used in reference to "packed tower aeration." These towers are very effective in removing volatile organic chemicals; reported removal efficiencies can reach as high as 100% (i.e., to nondetectable levels), but are typically in the range of 90-99% for the compounds normally encountered at gasoline-contaminated sites. Air stripping towers are also the most cost-

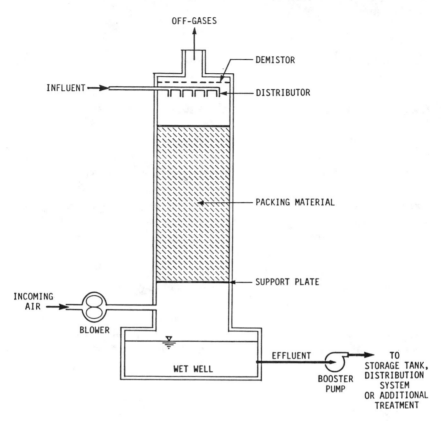

Figure 4.   Schematic diagram of packed tower aerator. *Source:* Camp Dresser & McKee, 1986.

effective of the stripping methods for most situations. For these reasons, this report will focus on countercurrent air stripping towers.

2.1.2   Theory of Air Stripping

The basic principles of air stripping are straightforward (Treybal, 1980). The kinetic theory of gases holds that molecules of dissolved gases can pass freely between the gaseous and liquid phases. At equilibrium, the same number of molecules move in both directions through a unit of area in a unit of time. The departure from equilibrium provides the driving force for mass transfer. The rate of mass transfer is proportional to the difference between the liquid-phase concentration of a contaminant in the bulk liquid and that substance's equilibrium liquid-phase concentration. The equilibrium concentration of a substance is dependent on its Henry's law constant. Henry's law describes the relative tendency for a substance to separate between the liquid and gaseous phases at equilibrium.

Thus, Henry's constant can be thought of as a partitioning coefficient. As will be seen later, the magnitude of Henry's constant is integral to the feasibility of air stripping for a particular compound. Henry's law can be expressed mathematically as:

$$p_a = HX_a \qquad (1)$$

where

p_a = partial vapor pressure of contaminant (atm)
H = Henry's law constant (atm)
X_a = mole fraction of contaminant a in water (mol/mol)

The phenomenon of air stripping can best be described as "controlled disequilibrium." Introducing fresh, contaminant-free air into the system results in a net mass transfer from the liquid phase to the gaseous phase. By continually replenishing the air with contaminant-free air, the contaminants are eventually reduced to very low levels. Air stripping with packed towers involves cascading water over randomly dumped packing while forcing air to flow countercurrent to the water. In this way, the effective surface area of liquid-air contact is greatly increased; hence the mass transfer is greater.

2.1.3 Design Parameters

The design of an air stripping column can also be described mathematically; the equations have been well-developed in the literature (Treybal, 1980; Kavanaugh and Trussell, 1981; Hand et al., 1986). The equations are derived by setting up a mass balance in the air stripper (Figure 5). Four basic assumptions are incorporated in these equations:

1. The influent air is free of volatile organic compounds (VOCs).

2. Plug flow conditions (i.e., where there is no differential flow) hold for the air and water flow. The use of an inlet water distribution system (Weir tray or nozzles) helps to preserve this condition.

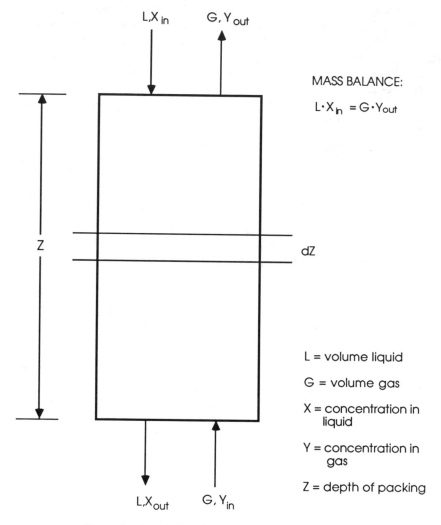

Figure 5. Differential element for an air stripping tower.

3. The changes that occur in the liquid and air volumes during mass transfer are negligible.

4. Henry's law holds true for these conditions.

To solve for the master design equation, two variables first need to be determined: the flow rate to be treated and the percent removal desired. The flow rate may depend on many factors, such as the extent of the contamination, the rate at which the contaminant plume is migrating, the future use of the water, the physical characteristics of the aquifer (permeability, transmissivity), or other factors. The desired removal efficiency is strongly dependent on the future use of the water, as well as the immediate health threat posed by the contamination.

The remaining design parameters can be determined once the flow rate and desired removal efficiency are known. It is important to realize that when designing a tower for a specific removal efficiency, a number of design parameters (e.g., size and type of packing, height and/or diameter of the column, water temperature, air-water ratio, gas pressure drop) can be adjusted to achieve similar results. Some of these, such as air-water ratio and tower height, are inversely related. The design objective of an air-stripping tower is to maximize the rate of contaminant removal from the water at the lowest reasonable cost. This is usually done by iterating various parameters to find the best combination. Chapter 3 describes in detail the design of a packed tower.

The required design parameters are Henry's constant (which is contaminant- and temperature-dependent), the mass transfer coefficient (which depends primarily on the packing material), and the stripping factor and air pressure drop, both of which are selected to minimize total cost while satisfying the removal efficiency goals.

Henry's Law Constant

Theoretical Henry's constants are available for most compounds of interest in the literature (ICF, 1985; Perry and Chilton, 1973). Figure 6 shows the Henry's constants for several gasoline constituents, among others. These values are computed from data on the compound's gram-molecular weight, water solubility data, temperature, and the equilibrium vapor pressure of pure liquid. There has been concern that Henry's constants derived from these values may not be correctly extrapolated to field design work. The low-solute concentration typical of groundwater, the temperature dependence of Henry's constant, and the fact that the inside of an air stripper does not represent true equilibrium are all reasons that have been given to doubt laboratory data. Recent work by Munz and Roberts (1987) has shown that solute concentrations do not affect Henry's constant, to at least concentrations as low as 0.001 M. Temperature was again shown to have a major effect on Henry's constant, and thus on stripper performance (Figure 7). Munz and Roberts stated that each rise of 10°C in tempera-

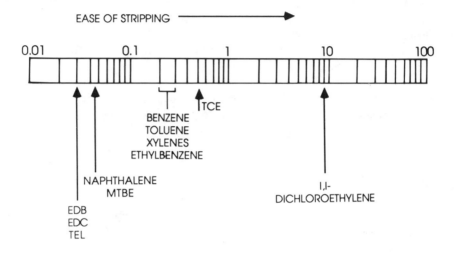

EDB = ETHYLENE DIBROMIDE (1,2 - DIBROMOETHANE)
EDC = ETHYLENE DICHLORIDE (1,2 - DICHLOROETHANE)
TEL = TETRAETHYL LEAD
MTBE = METHYL TERTIARY BUTYL ETHER
TCE = TRICHLOROETHYLENE

Figure 6. Range of Henry's law constants for compounds in gasoline and other compounds.

ture corresponded to an increase in the Henry's constant by a factor of 1.6. Thus, temperature is a very important consideration when designing a stripping tower.

Henry's law constants are typically expressed either as "dimensionless" or in the units of "atmospheres" (atm). Dimensionless units are valid only for systems that operate at standard pressure, since the actual units are:

$$\text{(atmospheres of pressure)} \frac{\text{m}^3 \text{ of water}}{\text{m}^3 \text{ of contaminant}}$$

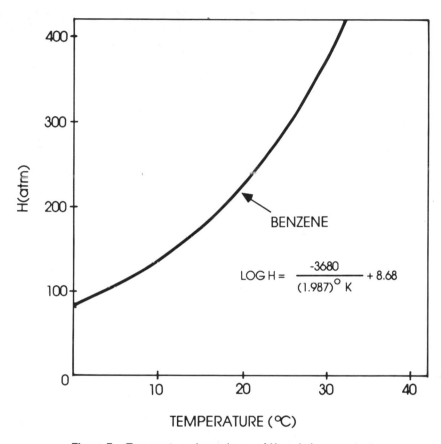

Figure 7. Temperature dependence of Henry's law constant.

Typical ranges of "dimensionless" Henry's constants for gasoline components are 0.02 to 0.30 (Figure 6). The more common unit is atmospheres, which is expressed by:

$$\text{(atmospheres of pressure)} \frac{\text{moles of water}}{\text{moles of contaminant}}$$

Typical values for gasoline components in atmospheres range from 20-500 (Figure 6). Care must always be taken when designing air strippers to use correct units for Henry's constant.

Mass Transfer Coefficient

The rate of mass transfer per unit time per unit volume is first order, proportional to the difference between the liquid-phase concentration in the bulk liquid and the equilibrium concentration:

$$J_A = - K_l a(C_i^* - C_i) \qquad (2)$$

where

J_A = the rate of mass transfer of contaminant A (kg/hr/m³)
$K_l a$ = the mass transfer coefficient (K_l) and the specific interfacial surface area (a); also known as the proportionality constant
C_i^* = equilibrium liquid-phase concentration (kg/m³)
C_i = operating liquid-phase concentration (kg/m³)

The proportionality constant, $K_l a$, is composed of the overall liquid mass transfer coefficient and the specific interfacial area. The mass transfer coefficient, K_l, represents the rate at which the system moves towards equilibrium (that is, the rate of mass transfer). The specific interfacial area, a, is a measure of the available total surface area of water exposed to the air. This value is dependent on the packing material. The best packing material will optimize the surface area per volume (m²/m³).

K_l is a function of the geometry and physical characteristics of the system, the compound being stripped, and the temperature and flow rate of the liquid. The two-phase resistance model of mass transfer is generally used to describe the kinetics of air stripping (Perry and Chilton, 1973). This model incorporates the resistance to mass transfer in both the liquid and gas phases. K_l is related to these by:

$$\frac{1}{K_l} = \frac{1}{k_l} + \frac{C_o}{k_g H} \qquad (3)$$

where

K_l = overall liquid mass transfer coefficient
k_l = liquid-phase diffusional resistance
k_g = gas-phase diffusional resistance
C_o = molar density of water (55.6 kmol/m³)
H = Henry's constant (atm)

When a compound has a large Henry's constant (above 50 atm), the term including k_g is negligible. In this case, $K_l \simeq k_l$. In our applications, it is valid to assume that the liquid-phase resistance dominates.

Values for $K_l a$ are sometimes supplied by manufacturers or may be found in the literature. However, because of the importance of this parameter in the tower design, it is recommended that $K_l a$ values be determined from pilot studies. In the absence of field data, there are two general methods by which these values can be determined. The first is the Sherwood-Hollaway empirical correlation:

$$\frac{K_l a}{D_w} = x \left[\frac{L'}{u_l}\right]^{1-n} \left[\frac{u_l}{p_l D_w}\right]^{0.5} \tag{4}$$

where

$$
\begin{aligned}
D_w &= \text{molecular diffusion coefficient in water (ft}^2\text{/hr)} \\
x, n &= \text{empirical constants} \\
L' &= \text{liquid mass flux rate (lb/ft}^2\text{/hr)} \\
u_l &= \text{viscosity of water} \\
p_l &= \text{density of water} \\
K_l a &= \text{units of sec}^{-1}
\end{aligned}
$$

A second, more common method is the Onda equations (Onda et al., 1968). These equations estimate the wetted surface area of the packing material and the liquid- and gas-phase mass transfer coefficients. These values are then used to obtain $K_l a$:

$$k_l = 0.0051 \left(\frac{u_l g}{p_l}\right)^{1/3} \left(\frac{L}{a_w u_l}\right)^{2/3} \left(\frac{u_l}{p_l D_w}\right)^{-0.5} \left(a_t d_s\right)^{0.4} \tag{5}$$

$$\frac{a_w}{a_t} = 1 - \exp\left(-1.45\left[\frac{T_c}{T_w}\right]^{0.75} \left[N_{Re}\right]^{0.1} \left[N_{Fr}\right]^{-0.05} \left[N_{We}\right]^{0.2}\right) \tag{6}$$

$$k_g = 5.23 \left(a_t D_a\right)\left(\frac{G}{a_t u_G}\right)^{0.7} \left(\frac{u_G}{p_G D_a}\right)^{1/3} \left(a_t d_s\right)^{-2} \tag{7}$$

where

$$
\begin{aligned}
N_{Re} &= \text{Reynolds number (dimensionless)} \\
&\quad \text{computed as}
\end{aligned}
$$

$$\frac{L}{a_t u_l}$$

N_{Fr} = Froude number (dimensionless) computed as

$$\frac{L^2 a_t}{p_l^2 g}$$

N_{We} = Weber number (dimensionless) computed as

$$\frac{L^2}{p_l T_w a_t}$$

a_w = wetted area of packing per unit volume (m²/m³)
a_t = total surface area of packing material per unit volume (obtained from manufacturer or literature, m²/m³)
k_l = liquid-phase mass transfer coefficient (m/sec)
k_g = air-phase transfer coefficient (m/sec)
T_c = critical surface tension of packing material (obtained from manufacturer or from the literature, kg/sec²)
T_w = surface tension of water (kg/sec²)
d_s = equivalent diameter of sphere with same surface area as a piece of packing material (m)
g = acceleration due to gravity (9.8 m/sec²)
R' = universal gas constant (0.0821) (L-atm/mol-°K)
L = liquid flow rate (kg/sec/m²)
G = gas flow rate (kg/sec/m²)
D_w = diffusivity in water (m²/sec)
D_a = diffusivity in air (m²/sec)
u_l = viscosity of water (kg-m/sec)
u_g = viscosity of air (kg-m/sec)
p_l = density of water (kg/m³)
p_g = density of air (kg/m³)

Several researchers (Hand et al., 1986; Wallman and Cummins, 1986) reported good agreement between $K_l a$ values derived from the Onda equations and pilot plant data. In general, the Onda-derived coefficients were somewhat lower than pilot plant data, and would result in a conservative design. An important conclusion by Wallman and Cummins (1986) was that $K_l a$ values increased with tower diameter.

This trend was attributed to sidewall effects, which were less important as tower diameter increased. Because of this finding, it was theorized that pilot plant determinations of $K_l a$ are also conservative.

Stripping Factor

The stripping factor, R, is a ratio of the actual operating air-water ratio to the theoretical minimum ratio. The minimum air-water ratio for 100% removal is determined by a mass balance in the stripper. It is based on the concept of Henry's law, which states that a certain amount of air must be brought into contact with the water to remove the contaminants. That minimum air-water ratio is described by:

$$(G/L)_{min} = \frac{C_i - C_e}{C_i} \times \frac{1}{H} \qquad (8)$$

where

$$(G/L)_{min} = \text{minimum air-water ratio}$$
$$H = \text{Henry's constant (dimensionless)}$$
$$C_i, C_e = \text{influent, effluent concentrations}$$

As described above, R is the ratio of the actual air-water ratio to this minimum ratio:

$$R = \frac{(G/L)_{actual}}{(G/L)_{min}} \qquad (9)$$

Combining these two equations by substitution, R can be expressed as:

$$R = (G/L)(H/P_t) \qquad (10)$$

where

$$P_t = \text{operating pressure } (= 1 \text{ atm})$$
$$H = \text{Henry's constant (atm)}$$

As can be seen, the stripping factor is directly related to the air-water ratio. In turn, these are related to the gas pressure drop through the column. There is more than one combination of air-water ratio and air pressure drop that will achieve a certain removal level. Therefore, these values are ultimately iterated to obtain the most cost-effective (considering both capital and O&M costs) design. Studies (Hand et al., 1986) have shown that the most cost-effective stripping factor (on a present-worth basis) usually falls between R = 3 and R = 5.

Gas Pressure Drop

The gas pressure drop through the stripping unit is usually found from a gas pressure drop curve. Many packing vendors will supply a brand-specific pressure drop curve; otherwise, a generalized curve may be used (Figures 8 and 9). Using this graph, it is possible to calculate the allowable gas and liquid flow rates for a variety of gas pressure drops. To use the pressure drop curve, find the appropriate value on

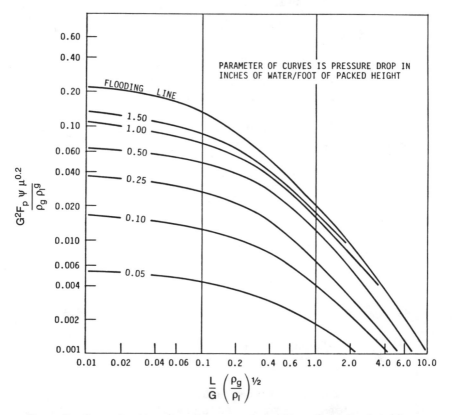

Figure 8. Generalized pressure drop curve for packings (English units). (From Eckert et al., 1970. Reproduced by permission of the American Institute of Chemical Engineers.)

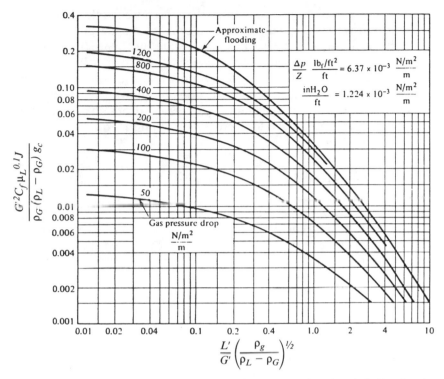

Figure 9. Generalized pressure drop curve for packings (metric units). (Reprinted from *Mass Transfer Operations* by R. E. Treybal, by permission. Copyright 1980, McGraw-Hill Book Company.)

the x-axis based on the selected air-water ratio. Read up to the chosen gas pressure drop (generally 0.25 to 0.50 in. of water/ft is used). It is usually better to use lower pressure drops for lower air-water ratios (Camp Dresser & McKee, 1986). By reflecting off the curve and reading the corresponding value on the y-axis, it is possible to calculate the allowable gas flow rate from the dimensionless group. Dividing the gas flow rate by the air-water ratio gives the liquid flow rate.

The pressure drop is a function of the gas and liquid flow rates and the size and type of the packing. It is important because it relates to the overall cost of the air stripper and the flexibility of stripper performance. A stripper operating at a high pressure drop will require a smaller volume than a similar stripper at a lower pressure drop. This reduces capital costs for the tower, but increases the blower cost. However, because the fan will be larger, more power will be required, thus increasing O&M costs. Various combinations of pressure drops and air-water ratios should be iterated to find the most cost-effective choice.

The pressure drop is also important because it relates to tower flexibility. Towers designed and built to operate at a low pressure drop have the flexibility to increase the gas flow rate, and hence the air-water ratio, should future influent concentrations increase or effluent limitations decrease. This ability will allow higher removal efficiencies, thus preserving present effluent concentrations or allowing attainment of stricter limits. Towers designed for high pressure drops do not have this capability, and would have to decrease the liquid loading to increase the air-water ratio.

2.1.4 Design Equations

After Henry's constant, the mass transfer rate coefficient, the stripping factor, and the gas pressure drop have been determined, all the variables of the master design equation are satisfied. This equation results from the solution of the mass-balance equation:

$$Z_t = \left(\frac{L}{K_l a}\right) \left(\frac{R}{R-1}\right) \ln\left[\frac{\left(\frac{C_i}{C_e}\right)\left(R-1\right)+1}{R}\right] \qquad (11)$$

where

Z_t = depth of packing (m)
L = liquid loading rate (m³/m²/sec)
$K_l a$ = overall liquid mass transfer coefficient (sec⁻¹)
R = stripping factor (dimensionless)
C_i = influent concentration (mg/L)
C_e = effluent concentration desired (mg/L)

This equation gives the total depth of packing necessary to reach the desired flow rate under the stated conditions. This can be thought of conceptually as:

$$Z = (NTU) \cdot (HTU) \qquad (12)$$

where

NTU = number of transfer units

$$= \left(\frac{R}{R-1}\right) \ln\left[\frac{\left(\frac{C_i}{C_e}\right)\left(R-1\right)+1}{R}\right] \qquad (13)$$

$$HTU = \text{height of transfer unit}$$

$$= L/K_l a \tag{14}$$

NTU, the number of theoretical transfer units, is a mathematical expression that characterizes the difficulty of removing a compound from solution. It bears a general relationship to the height of the stripping column. The value of NTU is predominantly influenced by the desired removal efficiency and to a lesser extent by the stripping factor (air-water ratio).

HTU, the height of a theoretical transfer unit, characterizes the rate of mass transfer from the liquid phase to the gas phase. The value is primarily influenced by the mass transfer coefficient and to a lesser extent by the liquid loading rate. The value bears a general relationship to tower diameter.

2.1.5 Design Procedure

There is no single procedure that must be followed when designing an air stripping tower. General procedures that can be used have been suggested in the literature (Kavanaugh and Trussell, 1981; Ball et al., 1984). Regardless of the procedure, values are first required for the flow rate, influent and effluent concentration, operating temperature, and Henry's constant for the limiting contaminant. After these initial values are determined, a suggested general design procedure is as follows:

1. Select the packing material. There are many commercially available packings, each with different mass transfer and pressure drop characteristics. Select a packing exhibiting a high mass transfer rate with a low gas pressure drop. For water treatment applications, plastic packings are most common because they offer low price, corrosion resistance, and lightweight (2-10 lb/ft^2) material that is easily dumped into a tower. Table 1 shows physical characteristics of common packing materials.

2. Select a reasonable stripping factor (between 2 and 10, with 3-5 being the best). Calculate the air-water ratio from Equation 10.

3. Refer to Figures 8 and 9. Select a reasonable gas pressure drop. Generally, it is better to choose lower pressure drops (defined as being less than or equal to 100 N/m/m^2) for low air-water ratios. Read the graph to find a value for the dimensionless group. Calculate the gas flow rate.

Table 1.   Physical Characteristics of Common Packing Materials

Type	Size	Surface Area (ft^2/ft^3)	Void Space (%)	Packing Factor[a] (1/ft)
Dumped Packings				
Glitsch	OA	106	89	60
Mini-rings	1A	60.3	92	30
(Plastic)	1	44	94	28
	2A	41	94	28
	2	29.5	95	15
	3A	24	95.5	12
Tellerettes	1″ (#1)	55	87	40
(Plastic)	2″ (2-R)	38	93	18
	3″ (3-R)	30	92	16
	3″ (2-K)	28	95	12
Intalox	1″	63	91	33
Saddles	2″	33	93	21
(Plastic)	3″	27	94	16
Pall rings	$5/_8$″	104	87	97
(Plastic)	1″	63	90	52
	$1^1/_2$″	39	91	40
	$2^1/_2$″	31	92	25
	$3^1/_2$″	26	92	16
Raschig rings	$^1/_2$″	111	63	580
(Ceramic)	$^3/_4$″	80	63	255
	1″	58	73	155
	$1^1/_2$″	38	71	95
	2″	28	74	65
	3″	19	78	37
Jaegar	1″	85	90	28
Tri-Packs	2″	48	93	16
(Plastic)	$3^1/_2$″	38	95	12
Stacked Packing				
Delta	–	90	98	–
(PVC)				
Flexipac	Type 1	170	91	33
(Plastic)	Type 2	75	93	22
	Type 3	41	96	16
	Type 4	21	98	9

Source: Taken from manufacturers' data and Treybal (1980).
[a]Represents "typical" value; actually a variable.

4. Based on the chosen air-water ratio, calculate the required liquid loading rate.

5. Find the tower diameter from

$$D = \left(\frac{4Q_{P_1}}{\pi L'}\right)^{0.5}$$

where Q = flow rate (ft^3/sec).

6. Find the height of transfer unit from Equation 14.

7. Find the number of transfer units from Equation 13.

8. Find depth of packing (Equation 12). Use an appropriate safety factor. (1.2 is common.)

9. Repeat for various values of stripping factor and gas pressure drop. Determine the most cost-effective combination of parameters based on present worth calculations. A detailed example problem is presented in Chapter 3 that illustrates how to design and cost out an air stripping tower.

2.1.6 Design Considerations

Several factors should be considered when designing an air stripping tower. One consideration is the character of the area surrounding the air stripper. If the area is residential, the tower, blower, and pumps may need to be enclosed for aesthetic reasons and/or to control noise levels. Depending on various factors (especially the gas flow rate), air strippers can be loud. Zoning laws may also affect stripper design. Many communities have maximum height limitations.

A second consideration is the prevailing wind patterns of the area. One of the assumptions of air stripping is that the influent air is free of VOCs. In order to ensure this condition, the air intake should be situated in such a manner as to prevent "short-circuiting" between the tower effluent air and influent air. Such a condition would result in lower removal efficiencies.

A third consideration is proper distribution of the influent water throughout the packing. A common problem is channeling along the wall of the tower ("sidewall effects"). This is caused by lower flow resistance there due to greater void volume. To correct this, redistribution of the water is provided by side wipers, normally placed every 20 ft of packing. In general, this problem is more severe with smaller-diameter columns.

A fourth factor to consider is the need for a gaseous demister.

This device captures any water entrained in the air prior to its exit to the atmosphere. These screens are fairly cheap ($200–300) and can prevent potentially significant quantities of water from leaving through the top of the column.

A fifth factor to consider is the material used for stripper construction with regard to the influent water quality. Aluminum is often used for construction because it is not susceptible to rusting. Fiberglass-reinforced plastic (FRP) or stainless steel could be used where water is especially aggressive. Resins used for FRP towers should be potable water/food grade and have EPA and FDA approval. Carbon steel is generally unacceptable due to its tendency to rust. If used, the steel should have potable water–grade coating. Concrete is sometimes used.

Other considerations, including efficiency problems associated with high iron/manganese content of the water and air pollution impacts, are addressed more fully under "Limitations" (Section 2.1.11).

2.1.7   Cost of Air Stripping

One of the main benefits of air stripping as a treatment technology for contaminated groundwater is its cost-effectiveness compared to other cleanup methods, such as activated carbon. However, the cost of air stripping is dependent on many factors and is therefore highly site-specific. Costs can vary widely, but air stripping is very often the least costly cleanup method for a particular site.

The total cost of any treatment method is a combination of the initial capital costs and the ongoing O&M costs. Capital costs are all costs associated with the startup of the air stripping facility. Included are costs for (1) the process equipment, such as the tower and packing material, air blowers, pumps, piping, valves, and electrical equipment; (2) a clearwell and holding tank (if needed); (3) any site-related costs, such as land purchase, bulldozing, and access; (4) vapor-phase control (if required); (5) materials and construction costs for housing (if required); and (6) miscellaneous costs such as painting, plumbing, and cleanup. Also included in the capital costs are fees for engineering and such contingencies as legal fees. O&M costs are basically comprised of (1) power for the pumps and blowers and (2) maintenance costs (labor and materials).

It is sometimes useful to determine the cost of treatment on a volume-treated basis. This is often done to compare the costs at different sites or to compare the costs of different types of treatment. A common expression used is the cost per 1000 gallons treated ($/1000 gal). This cost is basically the marginal cost of treatment. Typical treatment costs on a volume-treated basis are $0.05-0.25/1000 gal.

As described above, the total cost includes both capital and O&M costs. Determining the marginal O&M costs is fairly easy: divide the costs of power and maintenance for a certain period by the volume of flow treated in that period. Finding the marginal capital cost is more difficult; to do this, estimates must be made for the design life of the facility, the interest rate over that period, and the flow to be treated over the project life. The initial capital costs can then be annualized over the life of the project. Dividing by the estimated yearly flow will yield the marginal capital cost.

Cleanup costs at a particular site are a function of the length of the cleanup, the flow rate to be treated, the desired removal efficiency and/or the final concentration goal, the selected air-water ratio, the physical properties of the limiting contaminant, the residual concentration remaining in the aquifer, conventional water quality parameters, and the need for vapor-phase treatment, among other things. Each of these factors has a particular effect on the overall cost and the marginal cost of treatment. A summary of the various factors and their effects is given below.

Length of Cleanup

The length of cleanup can be one of the most important determinants of both total and marginal costs. A longer cleanup will usually mean higher initial capital costs but lower marginal costs. This is because the capital costs can be annualized over a greater number of years. The total operating costs will increase with time, but marginal operating costs are unaffected by the length of cleanup.

Flow Rate

The flow rate treated has a direct effect on the costs of treatment. A high flow rate will require a larger tower, clearwell, pumps, and blowers, and will require more electrical power than a low flow rate. Thus the total capital and O&M costs will increase with the flow treated. The marginal costs, however, will generally decrease as flow rate increases, due to economies of scale.

Desired Removal Efficiency

The desired removal efficiency and/or the final effluent goal have a primary influence on the total costs. In general, the higher the desired removal percentage (or the lower the effluent concentration limitation), the higher the capital and O&M costs. More complete contaminant removal (lower effluent concentrations) requires a higher air-water ratio, increased packing depth, or both (all other things being equal). Either factor increases capital costs, while a higher air-water ratio also increases operating costs.

Air-Water Ratio

The air-water ratio is a design parameter, chosen on the basis of cost-effectiveness and the limiting contaminant's Henry's constant. A higher ratio will increase power requirements but decrease tower volume. The engineer should determine the long-term price of higher operational costs versus higher initial costs, and choose this parameter based on the lowest present-value cost. For aromatic compounds, typical air-water ratios are 20-100:1.

Residual Concentration in the Aquifer

The contaminant concentration allowed to remain in the aquifer is an important cost consideration. As shown in Figure 10, costs are fairly constant for residual aquifer concentrations of 200-1000 ppb of total organic carbon (TOC). However, as the desired residual concentration approaches the low ppb range, costs increase exponentially. These total costs reflect the need for prolonged pumping life, reinjection of water to flush out the contaminants, and perhaps the use of detergents to loosen contaminants adsorbed to the soil particles. The residual concentration goal should depend on the future use of the site and the present danger of the contamination.

Conventional Parameters

The quality of the water in terms of traditional water quality parameters such as pH, hardness, and iron and manganese may affect the cost of any VOC treatment scheme. Abnormal pH, very hard water, and/or high levels of iron/manganese may require pretreatment of the influent prior to the air stripper. This could add considerably to the total cost.

Vapor-Phase Treatment

If treatment of the stripper offgas is desired or required, the total cost of stripping can be expected as a rule of thumb to double (Medlar, 1987). This assumption is based on the use of granular activated carbon (GAC) for treatment, and allows for the cost of the initial carbon

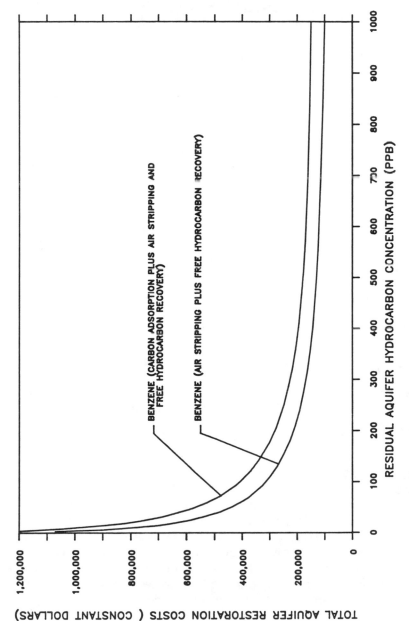

Figure 10. Total cleanup costs as function of residual aquifer concentration. *Source:* Engineering-Science, 1986.

charge, the contactor, and other site-related and construction costs. Vapor-phase treatment is discussed more fully in Section 2.1.12.

As can be seen, many factors influence the cost of air stripping. Because an air stripping tower can reach a certain removal efficiency through a variety of design parameters, an engineer should decide on the most cost-effective combination. To help with this complicated process, several computer cost models have been developed (Nirmalakhandan et al., 1987; Cummins and Westrick, 1982; Clark et al., 1984). Through the use of these models, it is possible to isolate one parameter and optimize costs based on this. For example, two studies optimized cost by iterating the gas pressure drop against the stripping factor. Another used the stripping factor and the liquid loading rate, while a fourth presented cost curves based on liquid flow rates for a variety of alternatives. From these studies, it appears that the most economical stripping factor is between 3 and 5.

To get actual cost figures, three sources were used: a survey of manufacturers and suppliers of packed tower equipment, case studies from published data that reported cost estimates and after-the-fact costs of cleanups, and cost curves developed by Camp Dresser & McKee (1987a). A survey of tower suppliers resulted in a range of costs from a low of $5000 (rated to treat 22 gpm) to $40,000 (rated to treat 450 gpm). These cost quotes generally include the tower, packing material, demister, blower fan and motor, and flow meter. The costs are primarily dependent on the rated flow rate, but are also influenced by "extras" such as sampling valves. Many suppliers custom-design towers for each particular case. Therefore, their costs varied more widely. Large strippers (rated over 500 gpm) were generally custom-built, and thus there are no quoted prices for towers this size. It is assumed that these cost proportionally more than the towers quoted above.

The survey of costs from cases reported in the literature yielded a range of capital costs from $27,000 to $1,100,000, and O&M costs from $7000 to $50,000 annually. According to these reports, the cost of the process equipment (tower, packing, pumps, and fans) accounted for between 20 and 75% of the overall capital cost, with higher numbers if air pollution control was required. The fees for engineering and contingencies normally ranged between 20 and 30% of the total capital costs. Where necessary, buildings and sitework contributed a significant part of the total cost of the facility (up to 50%). The wide range of costs exhibited can be attributed to the factors listed above, especially the flow rate, and to whether offgas pollution control is included. For example, the $27,000 case treated 70 gpm; the site costing $1,100,000 included five towers, each 12 ft in diameter and 50 ft high, which in combination treated 3500 gpm to drinking water levels.

Figures from Camp Dresser & McKee (Figures 11 through 16) give general capital and operating cost estimates for air strippers over a wide range of conditions.

Typical costs for air stripping towers at UST sites are about $130,000-150,000 (capital) and $6000-8000 annual O&M costs.

2.1.8   Removal Efficiencies

The ability of an air stripping tower to reduce VOCs to low levels has been demonstrated in hundreds of pilot-scale and full-scale operations. Like its cost, air stripping's achievable removal efficiency varies

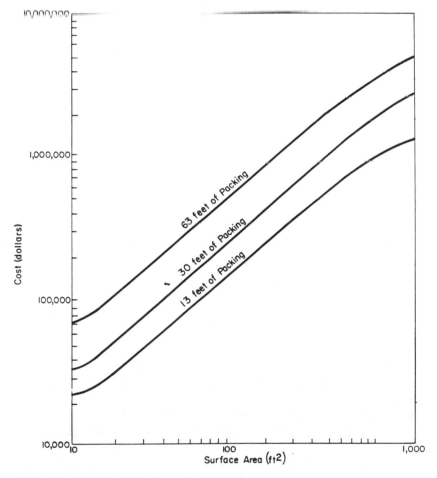

Figure 11.   Capital costs of a packed tower based on size. *Source:* Camp Dresser & McKee, 1987b.

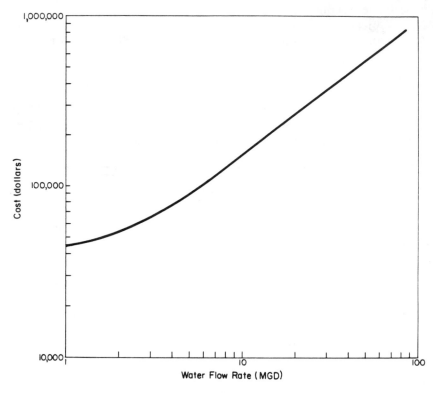

Figure 12. Capital cost of clearwell. *Source:* Camp Dresser & McKee, 1987b.

for differing sites and is influenced by a number of factors. Some of these are summarized below.

Water Temperature

The temperature of the influent water has a significant effect on removal efficiency, as shown in Figure 7. This is due to the temperature dependence of Henry's constant. (See Section 2.1.3.) Henry's constant increases with temperature (by a factor of ~1.6 per 10°C increase in temperature), resulting in higher rates of stripping for warmer groundwater. The temperature of groundwater is fairly constant throughout the year at a particular location, although it can vary in different areas of the country by as much as 15°C. This can have a strong bearing on the success of an air stripping facility.

Influent VOC Concentration

The influent contaminant concentration also affects the percentage removal. For similar conditions, a higher influent concentration will have a higher removal efficiency. This can be explained by recalling that the driving force for mass transfer is proportional to the difference between the operating concentration and the equilibrium concentration of the contaminant. As the operating concentration approaches the equilibrium concentration, the driving force decreases,

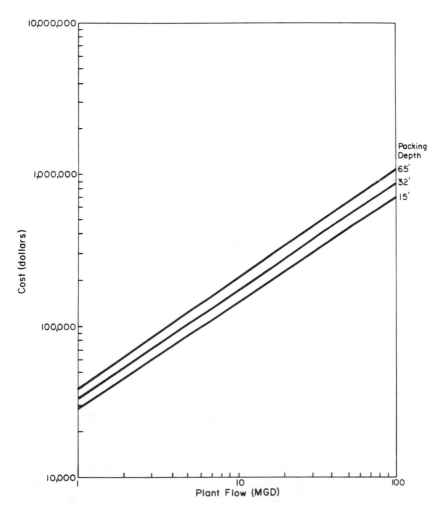

Figure 13.   Capital costs for water pump. *Source:* Camp Dresser & McKee, 1987b.

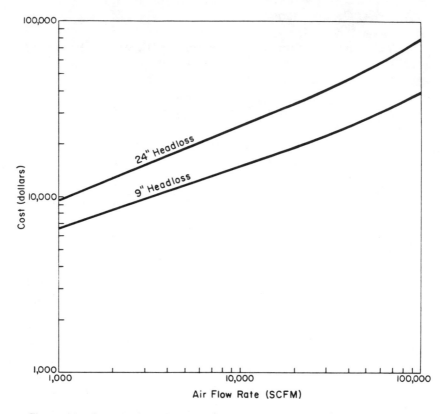

Figure 14.   Capital costs of air blower based on pressure drop. *Source:* Camp
Dresser & McKee, 1987b.

and relatively less contaminant is removed. For this reason, the final
effluent concentration as well as the percentage removal should be considered when judging a particular effluent goal.

Physical Properties of the Contaminants

The contaminants to be removed will influence the removal efficiency. This is due to the limiting compound's particular Henry's constant. Compounds with higher Henry's constants can be removed at a
higher percentage than those with lower Henry's constants. In cases
where multiple VOCs are present, the compound with the lowest
Henry's constant will generally be the limiting compound. However, a
compound at much higher concentration may be limiting even with a
higher Henry's constant.

Packing Material

The type of packing material can also affect the removal efficiency. The two broad categories of packing are randomly dumped

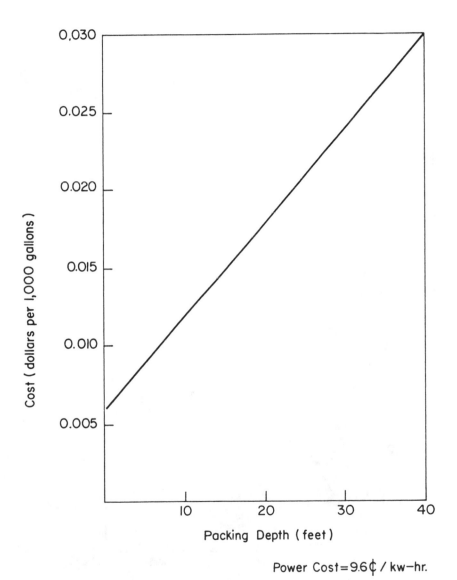

Power Cost=9.6¢ / kw–hr.

Figure 15.   Operating cost of pump based on packing depth. *Source:* Camp
Dresser & McKee, 1987b.

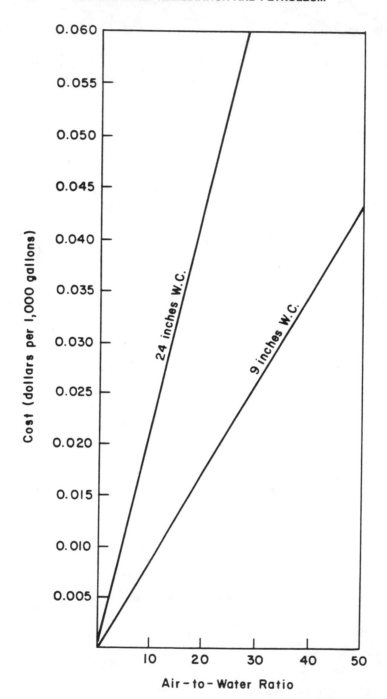

Figure 16. Operating costs for blower based on pressure drop. *Source:* Camp Dresser & McKee, 1987b.

Power Cost = 9.6 ¢ / kw-h

packing and stacked packing. Dumped packing uses small, randomly placed plastic, metal, or ceramic packings to provide a high surface area and a high void volume. Stacked packing can be described as performing like a "bundle of tubes." Dumped packing has been much more common, but stacked packings may offer some advantages. According to manufacturers, stacked packings are less susceptible to biological and mineral fouling. This is due to the higher (in some cases) void space and the fact that stacked packings do not have horizontal surfaces as do dumped packings. Table 1 gives physical characteristics of several common random and stacked packings.

Air-Water Ratio

Increasing the air-water ratio will usually result in increased removal efficiency. However, this effect may have diminishing marginal returns (Hand et al., 1986). For most gasoline compounds, very high-level (>99%) removal requires a very high air-water ratio.

Data from full-scale operations have shown that removal of 95–99% of the influent concentration of VOCs can normally be reached. In some cases, the product water is used for drinking water. Air stripping is most effective for removing low–molecular weight, nonpolar compounds with low solubilities; benzene, toluene, xylene, and other aromatics are normally removed to very low levels.

It is important to realize that the removal efficiency of an air stripping tower is fixed by the design, and will not change over the life of the cleanup (assuming initial conditions do not change). This differs from activated carbon, whose removal effectiveness depends on the life of the carbon and generally decreases over time for each carbon charge.

2.1.9 Ease of Operation

One of the main advantages of air stripping is its relative ease of operation. Once the tower, blower, pumps, valves, electrical instrumentation, and appurtenances are in place and operating, the facility is practically self-operating. There is no recurring maintenance (such as carbon replacement) requiring the services of an engineer beyond normal maintenance. However, iron and manganese or biological interferences may cause operational problems that would require the services of an engineer.

2.1.10 Reliability

The ability of air stripping to consistently produce high-level removal efficiencies for volatile groundwater contaminants is well documented. In the past, removal to low ppb levels or to below-minimum-detection-level (BMDL) of benzene, toluene, and xylene has normally been achieved. However, each site has its own characteristics and problems, and complicating factors may prevent achieving such low levels at every site. More conservative designs may add a safety margin for low-level removals.

2.1.11 Limitations

The use of air stripping for the removal of dissolved gasoline from groundwater may be limited by several factors. These include the types of chemicals that can effectively be removed by air stripping; possible air pollution impacts of the stripping tower effluent; high iron and manganese and/or suspended solids concentrations in the influent water; and possible high noise levels from the stripper.

Perhaps the most important limitation of air stripping is that many types of groundwater contaminants cannot be removed by this method. It is only applicable to the removal of volatile compounds. The major constituents of interest in gasoline, such as benzene, toluene, xylene, and ethylbenzene, are all fairly volatile, and thus easily removed. Compounds with low volatility, such as ethylene dichloride (EDC), are not readily removed by this technique. In general, very soluble compounds, high-polarity compounds, and high–molecular weight compounds are not easily removed by stripping.

The possibility of air pollution from the gaseous effluent from air stripping towers has caused concern. The operation of a stripping tower does not destroy the contaminant; it simply transfers the contaminant from the liquid to the gaseous phase. It is assumed that through the dilution occurring in the tower (due to the air-water ratio) and the natural mixing in the atmosphere, the ambient concentration of the contaminant in the surrounding air will be below safety thresholds. New Jersey, California, Michigan, and other states have regulations in place limiting the discharge of volatiles to the atmosphere. In New Jersey, a source is permitted to discharge no more than 0.1 lb/hr of any particular VOC, including benzene (Camp Dresser & McKee, 1987b). For strippers exceeding this limit, offgas air pollution control is required. Typically, carbon adsorption is used to treat the vapor-phase contaminant. Figure 17 shows some discharge rates at representative conditions. This graph shows the amount of a particular volatile

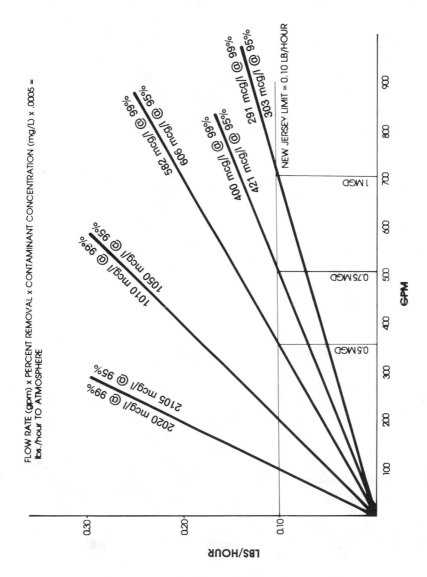

FLOW RATE (gpm) x PERCENT REMOVAL x CONTAMINANT CONCENTRATION (mg/L) x .0005 = lbs./hour TO ATMOSPHERE

2020 mcg/l @ 99%
2105 mcg/l @ 95%

1010 mcg/l @ 99%
1050 mcg/l @ 95%

582 mcg/l @ 99%
606 mcg/l @ 95%

400 mcg/l @ 99%
421 mcg/l @ 95%

291 mcg/l @ 99%
303 mcg/l @ 95%

NEW JERSEY LIMIT = 0.10 LB/HOUR

0.5 MGD   0.75 MGD   1 MGD

GPM

LBS/HOUR

0.10   0.20   0.30

100   200   300   400   500   600   700   800   900

Figure 17. Representative VOC discharge rates at various flow rates and removal rates.

or total volatiles that would be released to the air at the stated flow rates and removal efficiencies. (It is interesting to note that at gas stations, discharges have been measured at 10 lb/hr of VOCs.)

Another limitation on air stripping may be high noise levels resulting from tower operation. If the facility is in a residential neighborhood, the noise could be very disturbing, especially if the tower is being operated at a high gas loading rate. One solution is to surround the tower with walls extending above the tower. At one location where this was done, noise levels were significantly reduced (Camp Dresser & McKee, 1987b).

High concentrations of iron and manganese and/or suspended solids in the influent water can also limit the effectiveness of air stripping. Iron and manganese will facilitate the growth of bacteria on the packing, causing decreased mass transfer rates and higher gas pressure drops. The presence of toluene in the influent is thought to contribute to this effect (Abrams, 1987). Suspended solids can cause similar problems if they are trapped by the packing. Two methods have been used to remove precipitated iron and manganese. At one facility with high iron, the packing was periodically removed and placed in a tumbler, where the precipitate was broken off (Camp Dresser & McKee, 1987b). At other facilities, the packing has been washed with an acid solution or hydrogen peroxide to remove bacteria growing on the packing. Chlorination can also be used to control bacteria growth.

2.1.12 Offgas Air Pollution Control Systems

Possible air pollution from the operation of stripping towers, a major concern in some areas, is a potential limiting factor for the use of air stripping as a treatment technique. In cases where treatment of the stripper offgas is desired or required, vapor-phase granular activated carbon is the most common treatment. This method transfers the contaminant onto the GAC after it has volatilized from the liquid. Other treatment methods include incineration and catalytic oxidation.

The advantage of using vapor-phase GAC after a stripper (as compared to using liquid-phase GAC and foregoing the stripper) is in the greatly increased adsorption capacity of the GAC in the vapor phase. By transferring the contamination to the vapor phase (via air stripping) prior to removal by GAC, the carbon can adsorb much more contaminant, and the carbon will last much longer; thus, O&M costs are significantly reduced. For example, Zanitsch (1979) reported a vapor-phase adsorption capacity for toluene of 26% by weight (260 mg/g). This compares favorably with a liquid-phase capacity of 2.6% (26 mg/g) (Dobbs and Cohen, 1980). Depending on the chemical in question, the vapor-phase adsorption capacity can be from 3 to 20 times higher than the liquid-phase capacity (Medlar, 1987).

For vapor-phase GAC to be properly utilized, the offgas relative humidity must be reduced to below 50%. This can be done by using desiccants or by heating the air. If the relative humidity is not reduced, the capacity of the carbon is significantly reduced because the water molecules occupy adsorption sites preferentially. Another factor in the design of a vapor-phase GAC system is the approach velocity: it must be kept below 100 ft/min for effective adsorption.

The cost for vapor-phase GAC systems is typically $100,000 for single-bed units and $120,000 for dual-bed units (Camp Dresser & McKee, 1987b). These costs do not include the cost of the carbon or the operational cost. These costs are fairly constant over a range of treatment sites. Table 2 gives approximate relative cost ranges for several treatment alternatives.

2.2   ACTIVATED CARBON ADSORPTION

2.2.1   Background

Carbon has been used as an adsorbent for centuries; the ancient Hindus reportedly filtered their water with charcoal (Cheremisinoff and Ellerbusch, 1978). More recently, the beverage industry has used granular activated carbon for water treatment since the 1930s. Municipal water treatment facilities began choosing GAC for control of taste and odor problems in increasingly large numbers in the mid-1960s (Bright and Stenzel, 1985). Because GAC has the ability to remove a

Table 2.   Relative Cost Factors[a] for Treatment of Groundwater

Technique	Capital	O&M	O&M (RCRA)[b]
Air stripping	1 (assigned)	1 (assigned)	1
Air stripping and Vapor-phase GAC	2.0	3.0	4.0
Air stripping and Liquid-phase GAC	3.0	3.0	4.5
Air stripping and Liquid-phase and Vapor-phase GAC	4.0	5.0	7.5
Liquid GAC only	1.5	4.0	8.0

Source: Camp Dresser and McKee, 1987b.
[a]Cost factors indicated are relative to air stripping.
[b]Indicates cost if spent carbon must be treated as a hazardous waste under RCRA.

large variety of compounds (including organics) from water, its use has increased greatly over the past 20 years with the widespread occurrence of organic contamination (including gasoline spills), both in surface waters and groundwater. Along with air stripping, it is one of the two most common methods for treating groundwater contaminated by volatile organics including gasoline.

2.2.2 Adsorption Processes

Adsorption is a natural process by which molecules of a dissolved compound collect on and adhere to the surface of an adsorbent solid. Collection of the molecules on the surface can be due to chemical or physical forces. In both cases, adsorption occurs when the attractive forces at the carbon surface overcome the attractive forces of the liquid. Chemical adsorption is said to have occurred when the attraction is so strong at the carbon surface that a chemical compound is formed.

Physical adsorption is due to van der Waals forces, which are extremely weak bonds compared to chemical adsorption. Van der Waals forces, which are common to all matter, are thought to be the result of the motion of electrons. Molecules held by van der Waals forces are weakly adsorbed and can be removed by changing the solute concentration or by adding enough energy to overcome the bonds. This capacity for removal of certain molecules adsorbed on carbon (i.e., regeneration) and repeated reuse is what allows activated carbon adsorption to be a cost-effective technology. In environmental engineering applications, adsorption usually refers to physical adsorption.

The mass transfer of a solute from the bulk liquid to the carbon surface has three basic phases (Figure 18). First, bulk transport carries the solute (contaminant) among the carbon particles themselves. This type of transport is affected by the type of carbon and the liquid velocity. Second, film transport occurs; here the solute diffuses from the bulk liquid across the theoretical hydrodynamic layer surrounding the carbon particle. The rate of mass transfer across this layer is assumed to be dependent on the mass transfer coefficient k (Perry and Chilton, 1973). Third, the particle undergoes intraparticle transport throughout the carbon pores. This step can be divided into pore diffusion, surface diffusion, and micropore diffusion. The internal pores of activated carbon are classified based on size as micropores (10-1000 Å) or macropores (over 1000 Å) (Cheremisinoff and Ellerbusch, 1978). Pore diffusion describes the process where the solute is transported into and through the macropores. The only reaction occurring is adsorption on the macropore walls. Surface diffusion occurs when particles already adsorbed on the pore walls move further into the carbon particle.

Micropore diffusion is the transport mechanism in which adsorbate is carried into the micropores and reacts with the carbon walls.

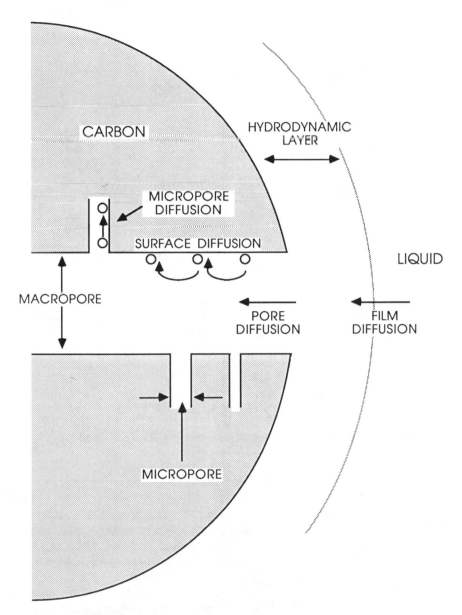

Figure 18.   Mass transfer of solute from liquid to carbon particle.

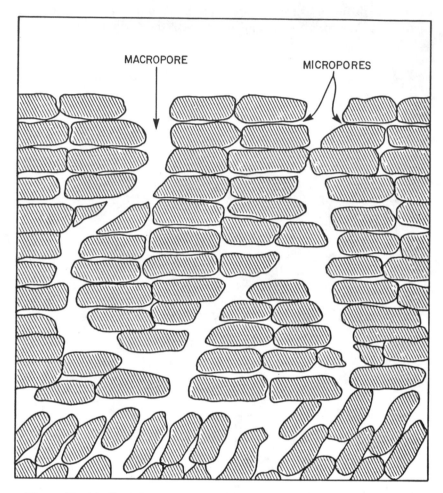

Figure 19.   Idealized diagram of internal pore structure of GAC. *Source:* Camp
Dresser & McKee, 1986.

2.2.3   Activated Carbon as an Adsorbent

Activated carbon is used as an adsorbent due to its large surface
area, a critical factor in the adsorption process. The typical range for
surface areas of commercially available activated carbon is 1000-1400
m^2/g. This very large surface area is due to the unique internal pore
structure of activated carbon (Figure 19). Most of the available surface
area is internal.

"Activated carbon" is actually a general term referring to a group
of substances. Activated carbon sold commercially originates from

several different sources, including bituminous coal, coconut shells, lignite, wood, tire scrap, and pulp residues, with coal the most common. To form GAC, the particular base is subjected to three steps: dehydration, carbonization, and activation. The dehydration step removes water by heating the material to 170°C. Increasing the temperature further drives off other vapors (CO_2, CO, CH_3COOH) and decomposition begins, resulting in carbonization. Activation occurs when superheated steam is released into the system, enlarging the pores by removing ashes produced during the carbonization step.

Activated carbon can be either powdered (PAC) or granular (GAC). Powdered carbon refers to particles whose size is less than U.S. Sieve Series No. 50; granular carbon is anything larger than this (Cheremisinoff and Ellerbusch, 1978). Powdered carbon is generally not recoverable in usable form. It is normally used as part of a treatment train, where it is added to the water and later removed by sedimentation or coagulation. Thus, PAC use is limited to complete treatment systems, where the product water is used for drinking water. Most leaking UST sites will not require extended treatment trains. Rather, GAC (which is normally recovered for reuse) is the usual choice at leaking UST sites when activated carbon is used. Therefore, this chapter will focus on granular activated carbon.

2.2.4 GAC Evaluation: The Isotherm

The basic instrument for the evaluation of activated carbon treatment is the adsorption isotherm. The isotherm is a function that relates the amount of solute adsorbed per weight of adsorbent to the solute concentration remaining in the liquid at equilibrium. As the term implies, isotherms are temperature-dependent, so values must be given in terms of temperature. The isotherm function, shown in Figure 20, can be thought of as a means of describing the capacity of carbon for a particular compound, or the efficiency of carbon to remove that compound.

The carbon capacity is influenced by a variety of factors: the solute to be adsorbed, the adsorbent (carbon) itself, the water temperature, the pH of the liquid, and other things. Isotherms are usually determined for a single-solute solution. If more than one compound is present in the water, as is usually the case at gasoline contamination sites, the isotherms are useful only for comparative purposes, rather than design purposes.

The equations most commonly used to describe experimental

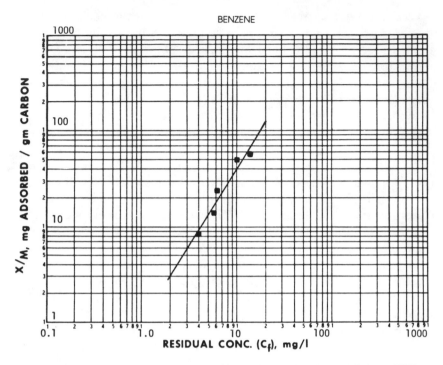

Figure 20. Freundlich isotherm for benzene. *Source:* Dobbs and Cohen, 1980.

isotherm data are those by Freundlich and Langmuir. The Langmuir isotherm is of the form:

$$\frac{X}{M} = \frac{QbC}{1 + bC} \qquad (15)$$

where

$\frac{X}{M}$ = amount of adsorbate (X) per weight of adsorbent (M)

Q = amount of adsorption per unit weight forming a complete monolayer

C = concentration of solute in water at equilibrium

$b = b_o \exp(-E/RT)$

where

b_o = a constant that includes the entropy term

E = energy of adsorption

R = universal gas constant
T = absolute temperature (°K)

The Langmuir isotherm equation was developed theoretically to closely model the adsorption process, as evidenced by the term Q, which assumes that a monolayer forms on the carbon. The more commonly used Freundlich isotherm, on the other hand, represents an empirical equation. It has the general form:

$$X/M = KC^{1/n} \qquad (16)$$

where

X/M = amount of adsorbate per weight of adsorbent
C = concentration of solute in water at equilibrium
K,n = empirical constants particular to the compound

The empirical constants K and n are determined by plotting experimental results on log-log paper, with the amount of solute adsorbed on the y-axis and the equilibrium solution concentration on the x-axis. The isotherm is typically linear. The slope of the line is equal to 1/n, while the y-intercept is equal to K. Although the constants have no physical significance, they are useful for comparing the adsorption capacities of different compounds or for testing the same compound on different carbons. (Isotherms are specific to the type of carbon used.) Values for these parameters are commonly found in the literature. Table 3 summarizes reported K values (which represent carbon capacities) for some gasoline components.

2.2.5 Activated Carbon Life and Breakthrough

Within an operating carbon tank (or bed), three distinct zones are present (Figure 21). The equilibrium zone, located at the influent end of the tank, is the area where the carbon's adsorptive capacity is exhausted (i.e., the carbon is saturated with contaminant). At the downstream end of the carbon tank is an area where the carbon contains its complete adsorptive capacity. Between these two zones is the mass transfer zone (MTZ), the area where adsorption is taking place. Within the MTZ a concentration gradient develops, with a high concentration at the influent end of the MTZ decreasing to near-zero concentrations for most contaminants at the downstream side of the MTZ. The length of this mass transfer zone depends on the loading rate and the charac-

Table 3.   Carbon Adsorption Capacities for Selected Compounds

Compound	Adsorption Capacity (mg/g)		Reference
Vinyl chloride		Trace	Nyer, 1985
Methylene chloride	Avg:	1.2	
		1.3	Camp Dresser & McKee, 1987b
		1.6	Camp Dresser & McKee, 1987b
		0.8	Nyer, 1985
		1.3	Dobbs & Cohen, 1980
1,2-Dichloroethane (EDC)[a]	Avg:	2.5	
		3.6	Dobbs & Cohen, 1980
		2.0	Nyer, 1985
		3.6	Hall & Mumford, 1987
		0.3	Hall & Mumford, 1987
Benzene[a]	Avg:	16	
		1.0	Dobbs & Cohen, 1980
		27.4	Camp Dresser & McKee, 1987b
		80	Verschueren, 1977
		4.1	Hall & Mumford, 1987
		1.73	Hall & Mumford, 1987
Ethylene dibromide (EDB)[a]	Avg:	17.0	
		17.0	Neulight, 1987
Toluene[a]	Avg:	22.5	
		26	Dobbs & Cohen, 1980
		50	Verschueren, 1977
		2	Hall & Mumford, 1987
		12	Hall & Mumford, 1987
Ethylbenzene[a]	Avg:	24	
		53	Dobbs & Cohen, 1980
		18	Verschueren, 1977
		2.2	Hall & Mumford, 1987
p-Xylene[a]	Avg:	46	
		85	Dobbs & Cohen, 1980
		55	Hall & Mumford, 1987
		50	Hall & Mumford, 1987
		28	Bright & Stenzel, 1985
		13	Bright & Stenzel, 1985
Naphthalene[a]	Avg:	68	
		132	Dobbs & Cohen, 1980
		5.6	Nyer, 1985
Phenol[a]	Avg:	91	
		161	Verschueren, 1977
		22	Dobbs & Cohen, 1980
bis(2-Ethylhexyl) phthalate		11,300	Dobbs & Cohen, 1980

[a]Gasoline components.

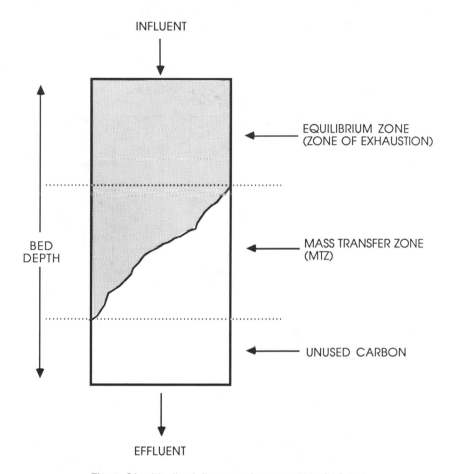

Figure 21. Idealized diagram of zones within GAC bed.

teristics of the adsorbent and adsorbate. The total length of the MTZ represents the resistance to adsorption.

The mass transfer zone moves down through the column as the total volume of water treated increases. Eventually, the leading edge of the MTZ reaches the end of the column (Figure 22), and the effluent contains increasingly higher concentrations of contamination as time passes. When the effluent concentration reaches a certain concentration (determined arbitrarily or based on effluent standards), breakthrough is said to have occurred, and the carbon is normally replaced. Figure 23 shows an idealized breakthrough curve. The breakthrough characteristics are an important determinant in deciding whether GAC

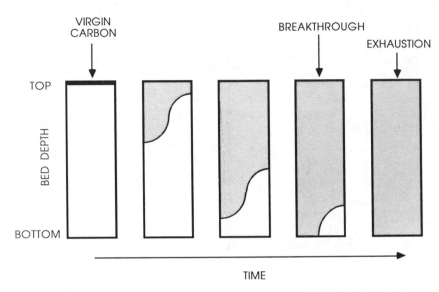

Figure 22. Breakthrough and exhaustion in an operating GAC bed. *Source:* Camp Dresser & McKee, 1986.

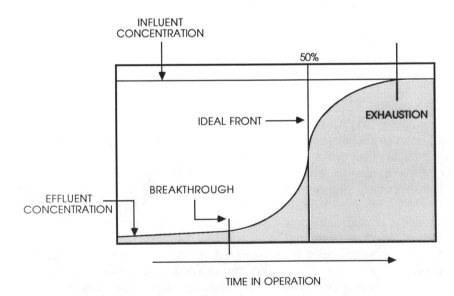

Figure 23. Idealized single-solute breakthrough curve.

is appropriate for a particular site. Breakthrough is discussed in more detail in Section 2.2.9.

2.2.6   Design of Carbon Systems

The design of an activated carbon system is not as straightforward as the design of an air stripping tower. Rather than the basic equations used to determine the size and operating parameters of a stripping tower, GAC design requires more complete pilot testing and judgment. The adsorption characteristics of any particular combination of contaminants are not generally predictable, except in a few situations where certain common chemicals are found and the engineer has vast experience. Even in these situations, a pilot test using the water of interest is many times required to accurately forecast the optimal empty bed contact time (EBCT) and carbon usage rate at a specific site.

When designing a GAC system, the EBCT is chosen first. The EBCT is defined as the volume of carbon divided by the flow rate. The EBCT relates directly to the size of the contactor needed; a high EBCT requires more carbon. The EBCT is inversely related to the carbon usage rate; the higher the EBCT, the lower the usage rate. The goal of the design of a GAC system is to find the optimal point in the tradeoff between a lower carbon usage rate and a smaller contactor size. A typical minimum EBCT for gasoline spills is 15 min. For a standard 20,000-lb supply of carbon in a 10-ft–diameter column, this EBCT results in a liquid loading rate of 2 gpm/ft^2. Experience has shown (Neulight, 1987) that this configuration results in a system that has a good removal rate and high flexibility, should future conditions change.

A second design variable is the decision to use a single-stage or multi-stage operation. This decision is based on the breakthrough characteristics of the influent stream, as well as on financial considerations. Influents exhibiting a long mass transfer zone are better operated in a multi-stage fashion (Figure 24). This mode allows a more efficient use of the carbon, although at a higher overall cost. The operation of a carbon system is discussed more fully in Section 2.2.7, below.

2.2.7   Operation of Carbon Systems

Facilities using GAC at UST contamination sites are normally operated as fixed bed facilities. Contactors may be either gravity or pressure filters, and may be single-stage or multi-stage. Each choice offers benefits for specific conditions.

INFLUENT

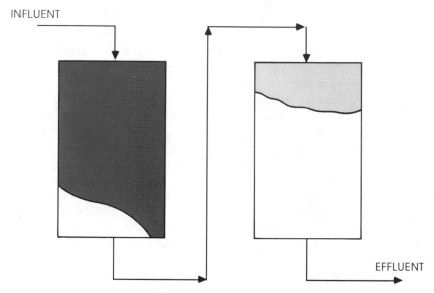

Figure 24. Schematic diagram of multi-stage GAC contactors.

EFFLUENT

    Fixed bed columns may employ upflow or downflow of the liquid. If downflow is used, the carbon bed acts as a filter for suspended solids in addition to removing organics. Filtering may be undesirable in some cases (for example, where the suspended solids concentration is high) because of the resulting high head losses and the necessity of extra backwashing; in these cases, upflow of the water would be preferred.

    Gravity filters are usually made of concrete and are operated similarly to sand filters. They are generally used for very high flows, such as are common at municipal water treatment plants (1-5 MGD). At most leaking UST sites, gravity filters are not used. Rather, pressure filters are used because of two major benefits: (1) they allow higher surface loading rates (5-7 gpm/ft² compared to 2-4 gpm/ft² for gravity); and (2) they allow pressure discharge to the distribution system, saving repumping costs. They are limited to diameters of 12 ft or less, due to the availability of the cylinders, and are normally 10 ft in diameter. A vessel 10 ft in diameter and 10 ft high holds approximately 20,000 lb of carbon. When wet, this weighs about 40,000 lb. This is the maximum allowable weight for shipment on U.S. highways, and thus is a convenient and typical size.

GAC contactors may be operated as either single-stage or multi-stage (in series). In multi-stage use (Figure 24), the leading contactor removes the majority of the contamination, while the second contactor acts as a "polishing" step, removing any residual organics from the water. In series operation, the entire adsorptive capacity of the carbon is used. The lead contactor can be used past breakthrough (to exhaustion) because the second contactor continues to remove the constituents. After replacing the spent carbon, the piping is reversed so that the new carbon becomes the polishing bed. Multi-stage operation is the optimal use of carbon. However, the cost for this method is higher than single-stage, and may not always be justified, especially where discharge limitations are not stringent.

2.2.8 Removal Efficiency

A large number of case studies have demonstrated the ability of activated carbon to remove a variety of compounds in gasoline to nondetectable levels (>99.99% removal). The effectiveness of GAC at a particular location depends on several factors, but primarily on the compounds to be removed. The suitability of GAC for a site depends primarily on cost, and how it is influenced by factors such as influent concentrations, effluent use (concentration limits), and the composition of the groundwater, and on available treatment alternatives to GAC. For example, GAC can almost always reduce gasoline-contaminated groundwater to <1 ppb of benzene. However, in cases where the influent concentration is very high, and/or the discharge requirements are not strict, air stripping (either alone or prior to GAC) may be a more cost-effective and appropriate means of removing the benzene. Factors that influence the choice of GAC for groundwater remediation are summarized below.

Effectiveness

Activated carbon has been used successfully to remove many gasoline compounds from water. Not every compound can be removed by activated carbon, however. GAC works best for low-solubility, high–molecular weight, nonpolar, branched compounds (Bourdeau, 1987). According to Brunotts et al. (1983), a compound's solubility in water is the key parameter in determining how well it will adsorb. Low-solubility compounds are adsorbed better than high-solubility compounds, all other things being equal. For this reason, alcohols, ketones, and ethers are poor adsorbers, while most solvents and pesticides are excellent adsorbers. High–molecular weight compounds adsorb better than lower–molecular weight compounds, perhaps because of the higher van der Waals forces they possess. However, extremely high–molecular weight compounds, such as sugars, do not adsorb at all. These compounds are not usually found in groundwater. GAC has a higher affinity for nonpolar compounds than for polar compounds, due to the surface chemistry of the carbon. The polarity of a compound depends on the chemical and physical structure of its molecules. Polar compounds behave more like ionic compounds, while nonpolar compounds are more neutral electrically. Most components of gasoline, particularly benzene, toluene, and xylene, are nonpolar. The molecular structure of a compound will also influence its ability to adsorb on GAC. Molecules that are branched or have attached functional groups, such as chlorine, fluorine, or nitrogen groups, adsorb well. Pesticides generally exhibit extremely high adsorbability, due in part to their complex molecular structure.

Other factors besides the characteristics of the contaminants influence the effectiveness of GAC treatment. They include the properties of the carbon, the temperature of the water, the iron and manganese concentration of the water, the EBCT used, the occurrence of desorption, and the activity of bacteria in the carbon bed.

As discussed earlier, GAC can originate from several different source materials and can be prepared ("activated") by a variety of methods. For these reasons, different carbons have different adsorptive capacities. The surface area of a carbon is the most important factor in determining its efficiency, because the amount of adsorption is directly proportional to this value. The surface chemistry of various carbons differs also, but this has a minor effect compared to surface area. Regenerated carbon also differs from virgin (unused) carbon. According to Bourdeau (1987), virgin carbon is normally used in cases where the effluent is to be used for drinking purposes. Reactivated carbon, which costs significantly less, is normally acceptable for sites where the effluent is discharged to surface water or groundwater.

The temperature of the water affects adsorption (Snoeyink, 1983). Isotherms are derived at a specific temperature. As the temperature increases, adsorptive capacity decreases. The effect of temperature in groundwater cases is minimal, as groundwater temperature is fairly constant throughout the year.

Groundwaters containing significant levels of iron and manganese (above 5 mg/L) must be treated to remove these compounds prior to GAC treatment. If the iron and manganese are not removed prior to treatment with GAC, they will precipitate onto the carbon, clog the pores, cause rapid head loss, and eventually prevent flow through the carbon.

The EBCT is defined as the volume of GAC divided by the flow rate to the column. It represents the theoretical time that the GAC is in contact with the water; however, the actual time of contact is about half of the EBCT, because the interparticle porosity of GAC is roughly 50%. The optimum EBCT is unique to each facility. It varies based on the type of carbon used, the contaminants being removed, bed depth, flow rate, and influent and effluent concentrations. Values for EBCT reported in the literature vary widely, from 3 to 2000 min. This wide range reflects the variety of situations to which GAC has been applied. A typical minimum EBCT for gasoline compounds is 15 min. This corresponds to a surface loading rate of 2 gpm/ft². Where less-readily-adsorbed compounds are present, a 30-min minimum EBCT is used. Pilot studies can be used to determine the optimum EBCT for a particular site.

Desorption may occur if there is a sudden decrease in the influent concentration. Desorption is the reverse process of adsorption. This can occur if the influent concentration drops significantly; previously adsorbed contaminant molecules may then desorb to maintain equilibrium in the solution. This can result in an effluent concentration that is higher than the influent concentration. Displacement may also occur if more strongly adsorbable contaminants appear in the influent. If this happens, previously adsorbed compounds may be displaced (Figure 25), resulting in higher concentrations for those compounds in the effluent than the influent.

Granular activated carbon beds are an excellent medium to support biological growth. Bacteria often prosper on the surface of the GAC. Once there, the bacteria are able to degrade certain compounds from the bulk liquid and the surface of the GAC (Speital and DiGiano, 1987). The occurrence of biodegradation has several benefits. Perhaps the most important benefit is the increased service life of the carbon. Degraded compounds do not occupy sorption sites, leaving those sites available for other molecules. Speital and DiGiano found that this reservoir of empty sorption sites may serve to dampen variations in the effluent, preventing slugs of higher-concentration effluent in periods of increased influent concentrations. Van der Kooij (1983) discusses possible negative aspects of biological growth on GAC, including the formation of endotoxins, high colony counts, and possible anaerobic conditions.

Appropriateness of Using GAC

After determining whether GAC could effectively remove the contaminants of concern at a leaking UST site, it must be determined

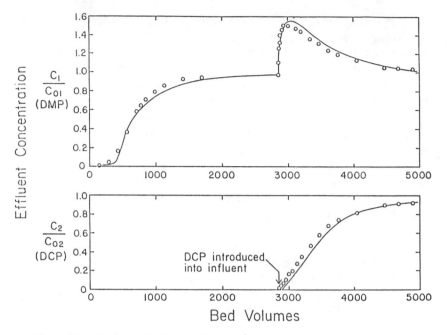

Figure 25. Displacement from GAC of dimethylphenol (DMP) by more strongly adsorbable dichlorophenol (DCP). *Source:* Snoeyink, 1983. (Reprinted from *Occurrence and Removal of Volatile Organic Chemicals from Drinking Water,* by permission. Copyright 1983, American Water Works Association Research Foundation.)

whether GAC is the most cost-effective technique to use. This decision will be based primarily on the influent concentrations of the contaminant(s) and TOC, the desired effluent concentration, and the breakthrough characteristics of the contaminants in the influent, because these are the primary factors affecting cost.

GAC is best suited for reducing low influent gasoline concentrations to nondetectable levels. In situations where the influent concentration of TOC is high, the carbon usage rates increase dramatically. O'Brien and Fisher (1983) reported the results of 31 contamination case studies (not all gasoline-related) where GAC was used. In 17 cases where the influent TOC was above 1 mg/L (1000 ppb), the median carbon usage rate was 1.54 lb/1000 gal treated. For the 14 cases where influent TOC was below 1000 ppb, the median carbon usage rate was 0.35 lb/1000 gal. Thus, treating a high influent concentration uses much more carbon and therefore is significantly more expensive than treating low influent concentrations.

Carbon is well suited to removing most gasoline contaminants to nondetectable levels. It is therefore an excellent choice where effluent standards are stringent, such as in the case of drinking water. Unlike air stripping, which has a specific percentage removal (of less than 100%), carbon removes compounds to nondetectable limits prior to breakthrough.

Waters with many contaminants (such as gasoline-contaminated waters) will increase the carbon usage rate significantly due to competitive adsorption. Conceptually, carbon has a limited number of adsorption sites. Each site can accommodate one molecule; once the site is filled, no other molecules can adsorb there. An influent with many compounds will have a carbon usage rate between that predicted by the compound of earliest breakthrough and that predicted by the sum of the usage rates of the individual compounds (Hall and Mumford, 1987). In some cases of competitive adsorption, displacement may occur if a more strongly adsorbable compound is introduced into the contaminant stream.

The breakthrough characteristics for each influent stream are also important in determining the appropriateness of GAC as a treatment technique. Section 2.2.9 (below) discusses breakthrough in detail.

2.2.9 Breakthrough

Breakthrough occurs when the adsorptive capacity of the carbon for a particular compound is exhausted, and that compound begins to appear in the effluent. Because each compound has a unique adsorptive capacity and because influent concentrations vary, compounds will "break through" at different rates.

The relative order of breakthrough of a group of compounds can usually be predicted based on the mean capacity of those compounds (X/M from isotherm studies). This is true in cases where the compounds have similar concentrations in the influent. Compounds with low capacities will be the first to appear in the effluent, whereas compounds with high capacities would likely appear later. Of the major components of gasoline, the general order of breakthrough (from earliest to latest) is benzene, ethylbenzene, toluene, xylene, naphthalene, and phenol (Figure 26). Other compounds sometimes found in gasoline, such as methyl-tertiary butyl ether (MTBE), 1,2-dichloroethane (EDC), and ethylene dibromide (EDB), might appear in the effluent even before benzene due to very low adsorption capacities (if the release is of one of the few gasolines containing these compounds). Table 3 gives the adsorptive capacities for several compounds, including gasoline constituents.

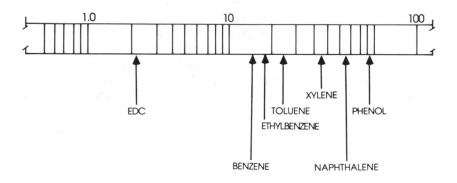

MEAN ADSORPTION CAPACITY, mg/gm
@ EQUILIBRIUM CONCENTRATION = 500 mcg/L*

*SOME EXPERIMENTS CONDUCTED AT 1000 mcg/L

Figure 26. Mean adsorption capacities of various compounds in gasoline.

In theory, breakthrough occurs when the edge of the mass transfer zone just reaches the end of the carbon bed, and effluent begins to contain a detectable amount of contamination (Figure 22). In practice, however, breakthrough usually refers to a point when the effluent reaches a certain level, or threshold, of contamination. This level is sometimes arbitrarily set, such as the commonly used 5% of the influent concentration (Reynolds, 1982); or the level may be based upon environmental regulations, such as discharge limits or drinking water standards. In either case, the level of contamination may refer to the total of all volatiles in the water (TOC is the usual indicator) or to a specific compound or compounds upon which the discharge limits are based.

For example, for an influent stream contaminated by a variety of compounds, to be used as a drinking water source, an effluent limit may be set for total volatile organics. This was the case at Acton, Massachusetts, which used GAC to treat municipal well water contaminated by several organics to levels of 150 ppb (MacLeod and Allan, 1983). There, an effluent standard of 5 ppb TOC was established for water for household use. Conversely, sometimes the effluent standard is based on a single contaminant. In these cases, the single contaminant may be that compound in the influent stream that is the first to break through, or it may be the compound considered most hazardous.

Usually, the compound with the earliest breakthrough is used as an indicator for carbon change. This is especially true in cases where the effluent is to be used for drinking water. In the case of gasoline contamination, benzene is normally the first compound to break through. For this reason, and because benzene is usually considered one of the most toxic components in gasoline, the effluent is typically monitored for benzene, and the carbon is changed when benzene reaches the threshold concentration. Note, however, that other compounds not found in all gasoline (such as MTBE and EDC) may break through earlier than benzene.

2.2.10   Cost of GAC Treatment

The cost of GAC treatment depends on site-specific conditions, and thus varies widely. *In general, though, GAC is more expensive than air stripping for similar situations.* This is because the capital costs of equipment and the 0&M costs for GAC are higher than those for air stripping.

Capital Costs

The capital costs of GAC treatment include the initial carbon charge, carbon vessel, the pumps and piping, electrical equipment, a clearwell (if necessary), housing (if necessary), and engineering, design, and contingencies. The flow rate and the discharge requirements are the conditions used for design and thus have a controlling effect on capital costs. For waters requiring removal to nondetectable levels, the carbon is normally operated in series. This requires two contactors, adding to the capital cost. Very high flow rates may be treated by using several pressure contactors in parallel, or by using a gravity contactor. Gravity contactors are often cement, and operate similarly to sand filtration tanks. They can accommodate surface loadings of only 2-4 gpm/ft^2 (Neulight, 1987) and thus must be larger than a corresponding pressure tank. Housing for the contactor(s) is often unnecessary. Engineering and contingencies average about 30% of the total capital cost. Figures 27-30 give approximate construction costs for four types of GAC contactors.

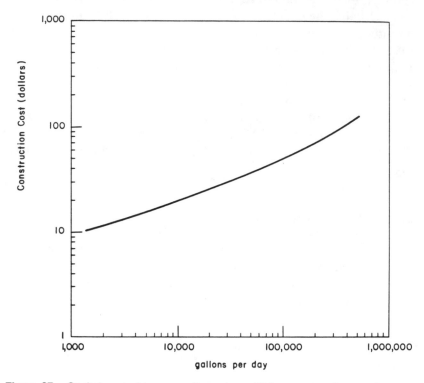

Figure 27. Capital cost of low-capacity package GAC contactor. *Source:* Camp Dresser & McKee, 1987b.

Operation and Maintenance

The operation and maintenance costs include the costs for labor, building upkeep, pumps and instrumentation (energy costs), carbon replacement/regeneration, and miscellaneous expenses. Labor, building upkeep, and miscellaneous costs would be similar to those for an air stripping facility. Where pressure filters are used, significant savings may occur because the water does not have to be repumped to the system. The cost of the carbon can be considerable, in many cases dominating the O&M costs or even the overall cost of treatment.

The carbon costs depend on the type of carbon used and the carbon usage rate. Carbon prices supplied by a manufacturer (Calgon) ranged from roughly $0.75 per pound for their highest (virgin) quality carbon to $0.60 per pound for service (regenerated) carbon. The carbon usage rate, a measure of how much carbon is required during treat-

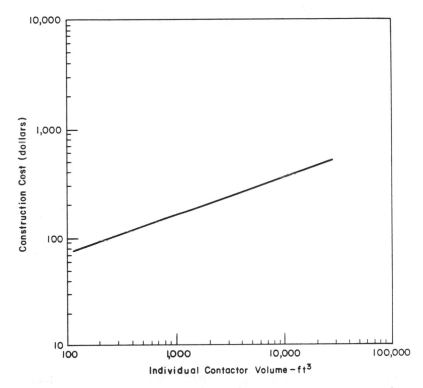

Figure 28.   Capital cost of pressure GAC contactor. *Source:* Camp Dresser & McKee, 1987b.

ment, is often expressed as pounds per 1000 gallons of water treated (lb/1000 gal). It is influenced by several factors, including the break-through characteristics of the contaminants to be removed, the concentration of contaminants, and the required effluent concentration.

As discussed in Section 2.2.8, each compound has a unique adsorption capacity that can be described by its Freundlich isotherm. The greater the X/M value (i.e., the weight of contaminants removed per weight of carbon), the lower the carbon usage rate. Studies (Hall and Mumford, 1987) have shown that the actual carbon usage rate for a typical water with several contaminants lies between the rate predicted by the compound with earliest breakthrough and the rate predicted by adding the usage rates of all the compounds. When designing a carbon system from theoretical isotherms, a large safety factor is normally used. Table 3 and Figure 26 give tabular and graphical representations, respectively, for adsorption capacities of gasoline components. If the

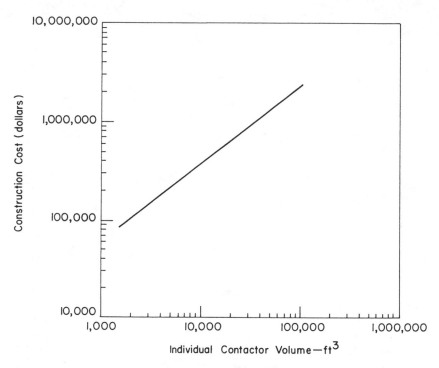

Figure 29. Capital cost of gravity steel GAC contactor. *Source:* Camp Dresser & McKee, 1987b.

spill contains MTBE, EDC, or EDB, which are sometimes found as additives to leaded gasoline, the carbon usage rate may be even higher.

The concentration of the contaminants in the influent stream has a direct effect on the carbon usage rate. As discussed earlier, the carbon usage rate increases dramatically with increasing levels of contamination in the influent. O'Brien and Fisher (1983) discussed 31 cases of contamination by various compounds where the carbon usage rate ranged from 0.1 to 13.3 lb/1000 gal. The high figures represent very high influent concentrations. A typical gasoline UST contamination site has influent concentrations in the range of 100-20,000 ppb TOC. This wide range takes into account both the size of the spill and the amount of dilution the gasoline has undergone in the aquifer. After cleanup has progressed for 6-12 months, the influent concentration often drops by an order of magnitude (Bourdeau, 1987) because the source has been removed and the treatment has removed much of the contaminant.

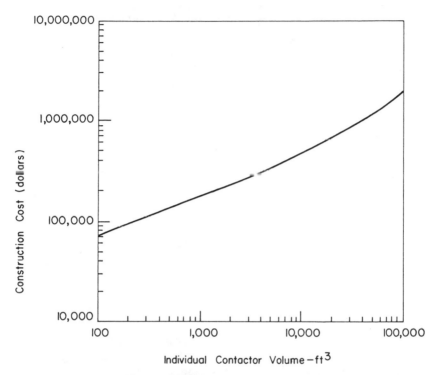

Figure 30. Capital cost of gravity concrete GAC contactor. *Source:* Camp Dresser & McKee, 1987b.

The discharge requirements for the effluent water also influence the cost. Water for potable use will have more stringent treatment requirements than water for groundwater recharge or surface discharge. These requirements may necessitate multi-stage treatment or a longer contact time, both of which will tend to increase overall costs. Multi-stage operation will actually decrease the carbon usage rate because the complete adsorption capacity is utilized, rather than the carbon being replaced at first breakthrough.

Because so many factors influence the cost of GAC, generalizations can be difficult. However, some treatment costs have been reported in case studies. As mentioned previously, costs differ widely depending on influent concentration, and reported figures are often segregated based on this variable. O'Brien and Fisher (1983) reported treatment costs on a per-volume basis. For influent concentrations above 1 mg/L, costs ranged from $0.45/1000 gal to $2.52/1000 gal.

Costs for lower concentrations (<1 mg/L) were between $0.22 and $0.54/1000 gal.

Total capital costs may range from $100,000 to $800,000, but normally fall around $350,000. O&M costs can range from $25,000 to $250,000 annually; typical figures are $25,000 to $40,000. Figures 27-30 may be used to find capital costs based on contactor size. Table 2 gives costs of GAC relative to air stripping.

Note that the relative cost factors in Table 2 are general rules of thumb, but may not be accurate in specific instances. For example, it may be less expensive to use air stripping and vapor-phase GAC than liquid-phase GAC where volatile concentrations exceed 100 ppb. This is because carbon usage is less for vapor-phase than for liquid-phase GAC. If volatile concentrations are less than 100 ppb, it may be less costly to use liquid-phase GAC than the air stripper and vapor-phase GAC. Capital costs for one air stripper and vapor-phase GAC contactors will be greater than for liquid-phase GAC contactors only. O&M costs for the liquid-phase GAC, however, will be greater. Initial contaminant concentrations and length of the cleanup ultimately determine which alternative is most cost-effective.

2.2.11 Reliability

Activated carbon has a long history in water treatment applications. It has been a proven means for removing dissolved organic compounds for over 15 years. During this time, GAC has been used to treat industrial wastewater and public water supplies, and as one of the main corrective action technologies for contaminated groundwater, including gasoline spills.

Activated carbon is generally very reliable for the removal of those compounds for which it is designed. GAC is an excellent technique for most organic chemicals found in gasoline, especially those with low solubilities. GAC is generally not suitable for highly soluble, highly polar, low–molecular weight compounds. These compounds either do not adsorb significantly or break through very early. Methanol, methylene chloride, and acetone are examples of compounds that are not readily removed.

Desorption is a possible phenomenon that may make GAC unreliable for certain treatment situations. This phenomenon (discussed in Section 2.2.8) occurs when influent concentration drops significantly or when new, more strongly adsorbable contaminants appear in the influent, and can result in higher effluent concentrations than influent concentrations for certain chemicals. Pilot plant studies should be made on a case-by-case basis regarding desorption.

2.2.12 Ease of Operation

The use of GAC requires a different type of supervision than stripping for effective operation. Because effluent quality decreases as time passes, the product water must be monitored regularly to ascertain when breakthrough occurs (unlike air stripping, which does not require constant supervision). At breakthrough, carbon replacement is necessary. As breakthrough becomes imminent, the system will require higher levels of attention. The replacement of carbon in the system requires an engineer and/or company representative to supervise the operation. Depending on the facility, replacement of the carbon can take from 1 to 12 hours, with pressure tanks requiring significantly shorter periods than gravity filters. It is advisable to have an engineer make regular inspections to make sure the facility is operating properly.

2.2.13 Limitations

The potential use of granular activated carbon for the removal of all dissolved gasoline constituents from groundwater may be limited by several factors. These factors include the adsorbability of the various components of gasoline, high iron and manganese content of the water, and disposal of the exhausted carbon.

Not every compound found in gasoline is amenable to adsorption by GAC. MTBE and diisopropyl ether (DIPE) are compounds sometimes found as additives to gasoline. Both of these, although they can be removed by GAC, have very high carbon usage rates (Garrett et al., 1986; McKinnon and Dyksen, 1984). The cost of removing these compounds by GAC is prohibitive, especially if the influent concentrations are substantial. For example, Rockaway Township, New Jersey found MTBE and DIPE in their drinking water at levels of 23 ppb and 14 ppb, respectively. They used GAC to remove these compounds and had to replace 40,000 lb of carbon every four weeks at a cost of $32,000 per replacement (McKinnon and Dyksen, 1984). Other compounds normally found in gasoline, such as benzene, toluene, xylene, ethylbenzene, EDB, and EDC, are all removable by GAC, with varying carbon usage rates (less than MTBE). Therefore, the presence/absence of highly soluble compounds such as MTBE and DIPE or other additives may determine the suitability of GAC for a particular gasoline spill. However, none of these compounds are believed to pose as significant a health concern as BTX in general, or benzene in particular. In addition, these additives are not found in all gasolines, unlike the BTX compounds, which are contained in over 99% of all gasolines. MTBE, for example, is found in only 10% of the gasoline being manufactured today (Garrett et al., 1986).

Iron and manganese levels in the influent water may also limit the use of GAC at a particular site. If these elements are present at levels above 5 mg/L, they must be removed prior to GAC treatment. If the iron and manganese are not removed, they will precipitate onto the carbon during treatment. If this happens, head losses will increase rapidly, the removal of organics will be hindered, and the carbon filter may eventually clog, rendering it ineffective. To use GAC at sites where iron and manganese are present at high levels, treatment to remove these elements to acceptable levels must precede the GAC unit. This may increase costs substantially, or be impractical due to space constraints.

A major potential limitation of GAC use is the disposition of the spent carbon. Usually, the spent carbon is either landfilled or regenerated. Regeneration is usually accomplished by heating the carbon to very high temperatures in a kiln to desorb the attached organics, and then incinerating the contaminants to destroy them. Regeneration can take place onsite or offsite at a central regeneration facility. Onsite regeneration is economically feasible only for the very largest projects; UST sites in general would not use this option.

Offsite (central) regeneration facilities have many limitations. After GAC is used to remove contaminants from water (or the vapor phase), it is laden with compounds and may be considered hazardous. For example, some spent carbon vessels may self-ignite; any carbon with a flash point below 200°F is considered hazardous, and may not be shipped over U.S. highways or accepted by a regeneration center. Likewise, most facilities will not accept carbon that has been used to remove dioxin or polychlorinated biphenyls (PCBs), due to possibly harmful air emissions. In addition, regeneration facilities have air effluent limitations, and thus may not accept all carbon for regeneration. Under RCRA, many contaminant-laden carbons may be considered hazardous materials, and thus need to be disposed of in a permitted landfill. This is the case for carbon used to remove tetraethyl lead (TEL), an additive to leaded gasoline. Because TEL precipitates onto the carbon, carbon contactor, and piping, the carbon and all equipment must be landfilled (as a hazardous waste). Quantities of carbon smaller than 20,000 lb are normally not accepted for regeneration.

2.2.14 Summary

GAC is an excellent technique to remove most organic compounds dissolved in water. Gasoline constituents, particularly benzene, toluene, and xylene, are normally reduced to nondetectable levels by GAC. This method is often more costly than air stripping, however, and may not always be the most cost-effective and appropriate method to clean up a gasoline spill. GAC use may be limited by site-specific conditions (such as high iron and manganese levels) or by the disposition of the spent carbon.

2.3   USING AIR STRIPPING AND GAC IN COMBINATION

2.3.1   Background

The previous sections established that air stripping and granular activated carbon are cost-effective techniques for organic chemical removal over a wide range of contaminant situations. In most situations involving gasoline-contaminated groundwater, either air stripping or GAC is the technique of choice. However, in some situations, the use of both air stripping and GAC is the best alternative on the basis of cost and organics removal.

The decision to use both methods in combination would normally be based on effluent quality and financial considerations. In all cases, the use of both measures in combination should result in an effluent quality as good as or better than use of either method alone. A phased approach is typically best suited for leaking UST situations. The first phase would consist of installing a packed air tower. Its performance could then be monitored to determine effluent concentrations and the need for additional (Phase Two) treatment with GAC.

2.3.2   Removal Effectiveness

In situations where effluent quality is required to be very high (such as potable water situations), the combination of air stripping and GAC is perhaps the best technique to reach nondetectable effluent contaminant levels. In these cases, air stripping is used first to remove a large percentage of the volatile organic contaminants, followed by GAC to reduce residual organic contaminants and any nonvolatile compounds to nondetectable levels. As seen in Figure 31, the use of air stripping as a pretreatment prior to GAC effectively puts an upper boundary on the effluent concentrations of VOCs (even at breakthrough) that is considerably lower than if air stripping were not used. Also, as the figure illustrates, GAC life is greatly extended. A properly designed air stripping tower can remove more than 95% of the volatile compounds from the influent. More importantly, many of the com-

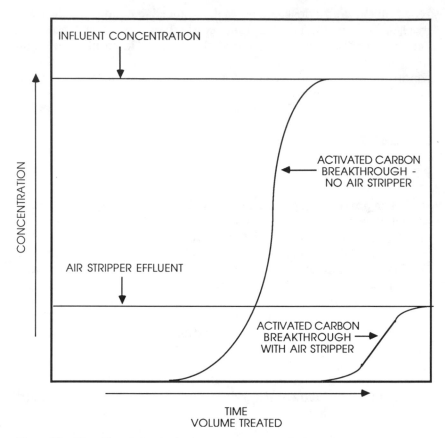

Figure 31.   The effect of air stripping as a pretreatment to GAC on effluent concentration.

pounds easily removed by stripping, such as benzene, methylene chloride, and dimethylamine, are those with the lowest carbon adsorption capacities. By removing these compounds, GAC will perform better and last longer, and effluent quality will be increased.

McIntyre et al. (1986), MacLeod and Allan (1983), and Camp Dresser & McKee (1986) have all reported the use of groundwater treatment systems with air stripping as a pretreatment prior to GAC. Influent to all the systems contained numerous compounds at varying influent levels. Effluent concentrations in all cases were below the detection limits for all the contaminants.

2.3.3   Cost-Effectiveness

In many cases, the cost of combining GAC and air stripping is the controlling decision variable. Because combined GAC and air stripping will nearly always yield higher-quality effluent than a single treatment method alone, it is safe to presume that the combination of methods would be used for any situation where this would result in a more cost-effective cleanup. A determination of capital and operating costs on a present-worth basis should be made for situations that might be suitable for the combination of air stripping and GAC.

Capital costs will increase if both air stripping and GAC are used. The cost advantage of using both methods together is found in decreased O&M costs (specifically, lower costs for carbon usage). More explicitly, adding a stripping tower prior to a GAC contactor is justified economically only if the savings on carbon replacement/regeneration equal or exceed the additional capital and O&M costs of the stripper. Stated simply, *if the total overall (capital and O&M) cost of treatment is lowered by adding a stripping tower, then that step should always be taken.* Conceptually, this is obvious. For an actual situation, determinations must be made of the decrease in carbon usage due to the air stripper and the associated cost savings. The decrease in carbon usage can be estimated based on the percent removal of the air stripping tower. A pilot test, either laboratory- or field-scale, using water from the site, should normally be performed.

As the positive slope of an isotherm shows, the adsorption capacity of carbon is greater at higher influent contaminant concentrations. Removing a portion of the volatile organics by stripping will lower the influent concentration, and the adsorption capacity of the carbon will decrease. However, the decrease in influent contaminant loading more than makes up for the decreased capacity, with the result being a lower carbon usage rate. O'Brien and Stenzel (1984) presented an example using trichloroethylene (TCE). Air stripping was assumed to have removed 80% of the influent concentration (1000 ppb), to 200 ppb. This lowered the adsorption capacity of the carbon from 57 mg/g to 27 mg/g. The carbon usage rate fell from 0.146 lb/1000 gal to 0.062 lb/1000 gal. Thus, an 80% decrease in TCE concentration via air stripping resulted in a 57.5% reduction in carbon use. Because of chemical similarities between TCE and BTX, comparable reductions could be expected for BTX.

2.3.4 Case Studies

Examples of actual cases where air stripping and GAC were used in combination are given below. They illustrate typical situations where air stripping was added after carbon regeneration costs became prohibitive.

Rockaway Township, New Jersey

In 1979, TCE was detected in the municipal wells of Rockaway Township, a small town in north central New Jersey (McKinnon and Dyksen, 1984). Subsequently, diisopropyl ether and methyl-tertiary butyl ether were also found in the water. The source of all three compounds was thought to be a leaking gasoline underground storage tank. Concentrations of TCE, DIPE, and MTBE in the water were 200–300 ppb, 70–100 ppb, and 25–40 ppb, respectively.

A decision was made to use GAC for water treatment. Based on initial estimates, the carbon supply (two 20,000-lb contactors) was expected to last six to eight months before replacement. However, after just three months of operation at a flow rate of 2 MGD, DIPE and MTBE had broken through and were measured in the effluent at levels of 14 ppb and 23 ppb, respectively.

The carbon was replaced at a cost of $32,000. Thereafter, carbon was replaced every two months. By the end of 1981, the carbon was being replaced every four to six weeks. (The annual O&M cost was about $200,000.) Therefore, a decision was made to add an air stripper prior to the GAC, at a capital cost of $375,000.

The stripper was sized to remove 99.9% of the influent DIPE concentrations, the least volatile compound (Camp Dresser & McKee, 1986). After the stripper became operational, the effluent was below detectable limits for all three chemicals. The GAC contactors were subsequently taken off line, as they were thought unnecessary. However, residents complained of a scaling problem in their hot water heaters. (The water in Rockaway Township is very hard.) It was thought that this problem was caused by the change in the water chemistry during stripping. Therefore, the GAC contactors were put back on line, and the scaling problem was solved. The carbon has not had to be replaced since installation of the air stripper.

Acton, Massachusetts

In December of 1978, two of the municipal wells in Acton, Massachusetts were taken out of service due to the presence of several organic chemicals, including trichloroethylene, benzene, and methylene chloride (MacLeod and Allen, 1983).

GAC was chosen to reduce the contamination from an average influent concentration of 42 ppb to less than 5 ppb total, and less than 1 ppb for any single compound (Nyer, 1985). The high cost of carbon replacement soon became prohibitive. Every five months, a complete replacement of 40,000 lb of GAC was required, at a cost of $37,000. In addition, influent concentrations were expected to rise and to cause even more frequent carbon changes. Therefore, a decision was made to use air stripping as a pretreatment.

The column was sized to handle 1 MGD at 95% removal of the VOCs. In practice, the removal ranged between 96-99% (to less than 1 ppb each), due to the safety factor used during design. The cost of the stripper was $31,000, and the building, electrical equipment, and miscellaneous equipment cost $109,000. Over the life of the project, this cost will be more than recovered through decreased carbon usage.

2.3.5 Summary

Using air stripping in combination with GAC is an excellent way to get very high-level removals of a wide range of compounds. For this reason, cleanup situations requiring high-quality effluent (such as to meet drinking water standards) may use both air stripping and GAC.

The use of air stripping and GAC may be justified in some cases as the least costly alternative. Such cases are likely to be those where carbon replacement costs are a significant portion of the overall cost of operation. Many times, these areas are those where influent concentrations are very high, and/or increasing, or where a contaminant in the influent breaks through very early. Two cases where these conditions were satisfied were given as examples.

2.4 BIORESTORATION

Indigenous, selectively adapted, or genetically altered microorganisms can potentially be used to degrade gasoline components dissolved in groundwater. The use of microbes to renovate contaminated aquifers is termed "biorestoration." This technique, although not yet as well known or as widely used as either air stripping or carbon adsorption, is a very promising method for groundwater cleanup. Unlike air stripping and GAC, which are separation techniques, biorestoration is a destruction technique. The end products of complete aerobic degradation are carbon dioxide and water. Also, where applicable, biorestoration is often the cheapest alternative available. Disadvantages of biorestoration include: (1) this technique cannot be used where a quick startup is needed (biorestoration typically takes 4-6 weeks for acclimation); and (2) it is not successful in a start/stop mode; that is, it must be continued 24 hours per day, 7 days a week.

Biorestoration can be accomplished in situ by either natural or induced methods. Natural in situ biorestoration occurs in aquifers as the microbial populations become acclimated to the pollutant and degrade the contaminants into simpler compounds (ultimately, carbon dioxide and water). Although the aquifer is used as the site of microbial activity as in natural systems, induced biorestoration makes use of systems to modify the groundwater regime to optimize degradation rates. Modification of the groundwater environment may be accomplished by use of various withdrawal, injection, and recirculation pumping systems to mix the contaminant with the media and microbial population; introduction of elements required for microbial growth, including oxygen, nitrogen, and phosphorus as well as growth substrates; or modification of the chemical characteristics (e.g., pH) of the groundwater to maximize microbial degradation rates. Biorestoration can also be effected by above-ground bioreactors constructed specifically to promote microbial growth in a vessel through which groundwater is pumped.

Regardless of which type of mechanical system is used for biorestoration, the fundamental processes of microbial degradation are essentially identical. This section provides a brief overview of the dynamics associated with microbial degradation of gasoline components and a review of the effectiveness, limitations, and costs associated with available biorestoration systems.

2.4.1 Microbial Processes

Microbial degradation of gasoline components can occur by aerobic respiration, anaerobic respiration, or fermentation. Aerobic microorganisms utilize oxygen in the process of decomposing hydrocarbons; anaerobes utilize inorganic compounds such as sulfate, nitrate, or carbon dioxide as terminal electron acceptors; and under fermenting conditions organic compounds serve as both electron donors and acceptors during microbe activity.

Major gasoline components such as the aromatics and alkanes, as well as some minor constituents such as ethylene dibromide (EDB) and ethylene dichloride (EDC), have been shown to be more readily degradable under aerobic than under either anaerobic or fermenting conditions. By-products of anaerobic decomposition, such as methane and sulfide, and of fermentation reactions, such as organic acids and alcohols, may also pose greater system management problems than those associated with the aerobic decomposition products carbon dioxide and water.

Although complete degradation of hydrocarbons will yield carbon dioxide and water, under certain environmental conditions complete degradation may not result. In such instances, intermediary degradation products that could be resistant to further degradation or inhibitory to further microbial growth may accumulate (Horvath, 1972).

Growth factors affecting the rate of microbial degradation include amount of oxygen, temperature, nutrient status, and growth substrate characteristics.

Oxygen Requirements

Aerobic degradation is the most attractive of the microbial processes for degradation of gasoline components in groundwaters because it proceeds at a more rapid rate and does not produce the noxious by-products associated with anaerobic decomposition. In order for aerobic degradation to occur, significant quantities of oxygen must be available to the microbes. Barker et al. (1987) calculated that 23.2 mg/L of oxygen is required for degradation of 1 mg/L of BTX in groundwater, and Wilson et al. (1986) noted that in a well-oxygenated groundwater containing 4 mg/L of molecular oxygen, microbes can degrade only 2 mg/L of benzene. The solubility of benzene in water (1780 mg/L) is therefore much greater than the capacity of microbes to degrade the compound under natural conditions. Because microbes will consume oxygen as the hydrocarbon is degraded, an aerobic groundwater can quickly became anaerobic. This onset of anaerobic conditions is the most significant factor in limiting the rate of biodegradation in the groundwater environment (Raymond, 1987).

Because of the importance of available oxygen in microbial degradation, this is the factor that would be most closely controlled when operating an in situ biodegradation cleanup. Three means of increasing the dissolved oxygen content of the groundwater are the injection of air, liquid oxygen, and hydrogen peroxide. According to Raymond (1987), the saturation concentration of oxygen in water from air injection is about 10 mg/L. Hydrogen peroxide injection can provide between 250-400 mg/L of dissolved oxygen. The very high amount of oxygen supplied by hydrogen peroxide makes it an excellent choice to maintain the aerobic condition of a groundwater system.

Temperature

Optimal growth of microbial populations responsible for biodegradation of petroleum products occurs from 20°C to 35°C. Microbial degradation rates would therefore be anticipated to be moderated by groundwater temperatures with decreasing rates occurring during winter months in northern portions of the country. However, experience has shown that biodegradation can occur at any groundwater temperature, once the microbes become acclimated.

Nutrients

Macronutrients such as nitrogen and phosphorus must be available for microorganisms in order to allow for biological processes to take place. The quantity of nutrients required for degradation is generally expressed as a ratio of the nutrient to the carbon source. For petroleum products the ideal C-N-P ratio is 160:1:0.08 (Bossert and Bartha, 1984).

Micronutrients such as magnesium and sulfur are also required for optimal growth, although in very small quantities. The micronutrients would not therefore be expected to limit growth of microbes in aquifer systems as often as oxygen deficiency. The specific nutrient requirements needed to optimize microbial degradation of gasoline components is a site-specific factor that must be determined experimentally for each groundwater contaminant problem. Part of the reason that determinations of nutrient requirements must be established on a case-by-case basis is because of the relationship of the substrate characteristics to the microbial populations.

Contaminant Type

The behavior of a mixed microbial population as encountered in a groundwater regime in reaction to the introduction of hydrocarbons will vary depending on the contaminant constituents and concentrations.

It has been demonstrated that bacteria, yeasts, fungi, or algae have the capacity to grow on straight-chain and branched alkanes (Singer and Finnerty, 1984), cyclic alkanes (Perry, 1984), and aromatic hydrocarbons (Cerniglia, 1984). Tabak et al. (1981) undertook investigations to determine the biodegradability of various organic contaminants using static culture enrichment techniques and wastewater microbiota. Table 4 summarizes the results of these 7-day screening tests at contaminant concentration levels of 5 and 10 mg/L for selected compounds.

Table 4. Microbial Degradation Screening Test Results

Compound	Performance[a]	
	5 mg/L	10 mg/L
Benzene	D	D
Ethylbenzene	D	A
Toluene	D	D
Phenol	D	D
Naphthalene	D	D
1,2-Dichloroethane	B	B

Source: Tabak et al., 1981.
[a]Performance at noted concentrations: A = significant degradation with gradual adaption; B = slow to moderate biodegradation concomitant with significant rate of volatilization; D = significant degradation/rapid adaption.

EPA is currently undertaking a comprehensive research program aimed at determining the land treatability of hazardous wastes (Matthews, 1987). The biodegradation of various gasoline components is being investigated:

- benzene
- toluene
- xylene
- phenol
- tetraethyl lead
- ethylene dibromide
- 1,2-dichloroethane

Component Concentration

Alexander (1985) reported that the rates of mineralization of some organic compounds are directly proportional to their concentration, and that there is a threshold level below which certain compounds usually subject to biodegradation are not converted to CO_2 and H_2O. At the higher concentrations of hydrocarbons in groundwater, such as would occur directly beneath a floating oil slick, microbial toxicity may occur (Cooney, 1984). As the concentrations of contaminants decrease and microbial populations become adapted to the compounds, the microbes may be able to overcome the effects of toxicity and degrade the compounds. The resilience of microbial populations to service and repopulate areas in which toxic levels of contaminants were initially present has been demonstrated repeatedly. Therefore, it is anticipated that biorestoration techniques could be adapted to deal with initially high contaminant levels. The more difficult question to address with respect to biorestoration is the level to which a contaminated aquifer can be cleaned up.

Microbial populations have been found to be higher in areas where hydrocarbon concentrations are high. Cooney (1984) reported that the rate of microbial degradation may be slow or nil below a certain threshold level. At low substrate concentrations, other mechanisms may be more significant in the reduction of contaminant levels than typical aerobic mechanisms. Schmidt and Alexander (1985) reported on work in which the effects of low levels of organic carbon on the degradation of low concentrations of substrate were evaluated. The study demonstrated that the rate and extent of biodegradation of low concentrations of synthetic organic compounds may be controlled by the presence of other organic molecules in the system. Pure cultures of bacteria were shown to utilize low levels of aromatics simultaneously in the presence of other organic compounds. The addition of organic substrates in biorestoration systems could therefore theoretically serve to further enhance biodegradation of hydrocarbons in groundwater at low levels.

Degradation of compounds may also occur in situations wherein the microbe effecting degradation does not derive any nutrients or energy from the process. This process, co-metabolism, is defined by Horvath (1972) as any oxidation of substances that takes place without use of the energy derived to support microbial growth. Co-metabolism may be a significant process for degradation of low levels of hydrocarbon contaminants in groundwater and should be evaluated on a site-specific basis when biorestoration techniques are being considered, since gasoline components such as xylene and ethylbenzene have been demonstrated to be subject to co-metabolism (Horvath, 1972).

Bouwer and McCarty (1984) showed that trace levels of the aromatic hydrocarbons ethylbenzene, styrene, and naphthalene could be reduced significantly in the presence of a primary substrate. Specifically, with acetate as a primary substrate at concentrations of 1.0 mg/L, ethylbenzene levels were reduced from 9.1 µg/L to 0.1 µg/L. Ethylbenzene present alone at these low concentrations could not trigger biodegradation, whereas the presence of acetate as a primary substrate was effective in stimulating degradation.

2.4.2 Effectiveness

Laboratory studies have shown biorestoration to be significantly effective. However, actual field applications of biorestoration techniques have pointed out the site-specificity and variable nature of these processes.

Amdurer et al. (1986) have summarized case histories of in situ treatment techniques applied to subsurface contaminants. Among the case histories summarized is that of the Biocraft site in Waldich, New Jersey. At this facility, an estimated 30,000 gal of methylene chloride, acetone, n-butyl alcohol, and dimethyl aniline leaked from underground storage tanks to subsurface soils and groundwater. The biorestoration system constructed for use at this site was comprised of a downgradient dewatering trench and well, two mobile biological activating tanks, two mobile settling tanks, and two upgradient reinjection trenches (Figure 32). The system was used to treat 14,000-20,000 gal/day of groundwater.

The median contaminant mass reduction ranged from 88 to 98%, with each pass through the system having a retention time of 12 hr, except for dimethyl aniline, where only 64% of the mass was reduced. Operation of the system was initiated in 1981 and as of 1985, 95% of the contaminants in the groundwater had been removed.

Yaniga (1982) reported on the biorestoration of a groundwater aquifer in Montgomery County, Pennsylvania, where gasoline had leaked from an underground storage tank at a service station. The contaminant plume initially contained up to 15 ppm of dissolved hydrocarbons. The biorestoration system utilized at this site consisted of a central pumping well and a reinjection gallery situated over the former underground storage tank pit. The extracted groundwater was passed through an air stripping tower to remove volatile organics and then oxygenated and enriched with nutrients prior to being reinjected in a batch process. The hydrocarbon concentration in the groundwater aquifer was reduced to 2.5 ppm after 20 months of system operation.

Yaniga and Smith (1985 and 1986) reported additional information on the above site following three years of the abatement program. They indicated that 30 to 3500 gpd of oxygenated and nutrient-rich water was injected into the infiltration gallery and after the first 11 months of operation the organic content of the groundwater was reduced by 50-85%. The authors report that hydrogen peroxide is superior to use of air in oxygenating the groundwaters. Although others have reported that H_2O_2 can cause microbial toxicity, Yaniga and Smith (1986) indicated that hydrogen peroxide at a 100-ppm concentration did not kill but enhanced microbial growth and yielded a 50-ppm oxygen concentration in the groundwater.

Barker et al. (1987) undertook investigations to evaluate the fate of BTX in a shallow sand aquifer by intensively monitoring the aquifer

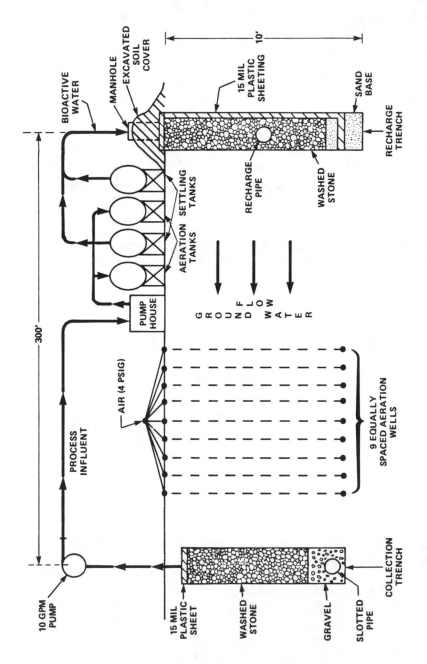

Figure 32. Biocraft biorestoration system flow diagram. *Source:* Amdurer et al., 1986.

following the injection of 1800 L of water containing about 7.6 mg/L o BTX. The attenuation of these components was evaluated. The field studies were compared to microcosm studies in which conditions approximating those encountered in the field were maintained.

The study conclusions support the general finding that oxygen is rate-limiting with regard to microbial degradation of petroleum products in the groundwater regime. BTX components were found to persist in aquifer layers where dissolved oxygen levels were near zero. Within about 1.2 years of BTX injection, complete natural removal was observed in the field experiment, with benzene exhibiting the greatest resistance to breakdown.

Chan and Ford (1986) reported on the use of both in situ and bioreactor systems to remediate a 22,000-ft^2 site contaminated with No. 2 diesel fuel. The estimated removal efficiency of the combined system following 100 days of operation was 80%. The authors noted that the bioreactor was 16 times more efficient than the in situ process in removing contaminants from the groundwater. The increased efficiency was attributed to the higher oxygen concentrations that could be maintained in the bioreactor (9 mg/L) as compared to the in situ levels (2.5 mg/L).

Dorr Oliver Incorporated has developed bioreactor systems, OXITRON® and MARS™, that could potentially be used for the biorestoration of groundwater contaminated with gasoline (Sutton, 1986). The OXITRON® system is an aerobic or anoxic fluidized bed system that can be used with sand, activated carbon, or other media on which biomass buildup occurs (Figure 33). Contaminated groundwater is passed through the system from the bottom of the reactor vessel at sufficient velocity to fluidize the bed. Because of the increased space between fluidized particles as compared to suspended growth media systems, five to ten times greater biomass concentration is reportedly achieved (Sutton, 1986). The increased biomass concentration reduces the hydraulic retention time required for treatment.

MARS™ (Membrane Aerobic or Anaerobic Reactor System) employs a well-mixed suspended growth reactor and ultrafiltration system that can be used for biorestoration of groundwater (Figure 34).

Although site-specific evaluations of the groundwater constituents' degradability in the OXITRON® or MARS™ system are needed (as with all bioreactor or in situ techniques), Dorr Oliver has reportedly used these systems for treatment of various wastes similar to those encountered in gasoline-contaminated groundwaters (i.e., benzene and toluene) successfully.

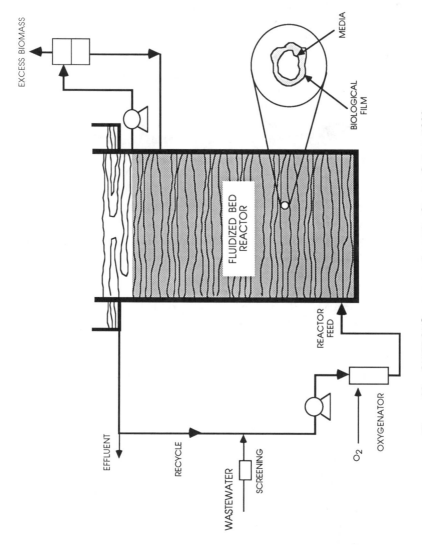

Figure 33. OXITRON® process schematic. *Source:* Sutton, 1986.

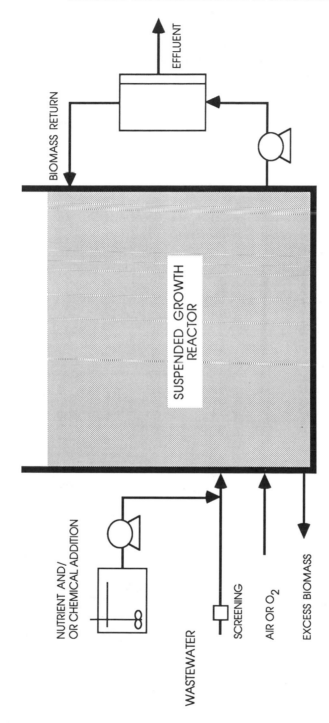

Figure 34. MARS™ process schematic. *Source:* Sutton, 1986.

Although indigenous microbial populations have the capability to degrade hydrocarbons, interest in use of mutant, genetically engineered, or laboratory adapted organisms has increased in recent years. Polybac Corporation of Allentown, Pennsylvania and Solmar Corporation of Orange, California are two enterprises providing microorganisms for use in biorestoration.

Polybac Corporation (1980, 1981) markets Polybac and Hydrobac, described as formulations of mutant, adapted microbes and biochemical accelerators. Polybac is made specifically for municipal and food processing wastewater treatment systems, while Hydrobac is formulated for petroleum refinery/petrochemical plant wastewaters. Other products available from Polybac Corporation include Petrobac for degrading hydrocarbons in salt water and Phenobac for degrading hydrocarbons in fresh water. Polybac Corporation also markets bioreactor systems and services (CTX-BIOX System) that have reported "removal efficiencies in excess of 99% for most organic compounds."

Solmar Corporation markets microbial cultures for use principally in wastewater collection and treatment systems. One of the products Solmar offers is Advanced BioCulture Formulation L-104, used to deal with "heavy, tarry types of oils, coal tars and organic sludges." It is reportedly well suited for aromatic and phenolic wastes (Solmar, 1974).

Other firms providing groundwater biorestoration services and systems include:

- Groundwater Technology, Chadds Ford, Pennsylvania

- Groundwater Decontamination Systems, Inc., Paramus, New Jersey

- NEPCCO, Foxboro, Massachusetts

- Detox, Dayton, Ohio

- Emtek, Bedford, New Hampshire

- TerraVac, Inc., San Juan, Puerto Rico

2.4.3 Limitations

Limitations on the use of biorestoration techniques include those related to sociopolitical issues and those associated with technical factors. Because in situ treatment is accomplished largely underground, little if any evidence of activity may be discernible by the general public. This lack of observable activity may lead to a public perception of no action. The public in general may be more prone to respond positively to corrective actions that manifest significant levels of activity, such as air stripping. Therefore, public relations efforts aimed at educating individuals about the mode and effectiveness of in situ degradation may be required as part of in situ degradation programs where high levels of public awareness exist.

Technical limitations on use of biorestoration techniques may be related more closely to site-specific characteristics than to the overall theoretical ability of microorganisms to degrade gasoline components in groundwater. As discussed, microbial growth factors appear to be readily modified either in situ or by use of bioreactors to allow for degradation of gasoline in groundwater systems. Degradation of hydrocarbons to ppm levels should be relatively attainable; however, treatment to ppb levels may require the manipulation of the system to encourage co-metabolism or degradation in the presence of an added primary substrate. These limitations can only be assessed by performance of site-specific laboratory- and pilot-scale evaluations. It is possible to meet the proposed maximum contaminant level (MCL) of 5 µg/L for benzene and 0.44 mg/L for xylene with a system that is appropriately designed and optimally operated.

The ability of biorestoration systems to be implemented in saline as well as fresh groundwater systems makes it adaptable to situations where other systems may not be suitable. The ability of the recovery system to capture groundwaters contaminated above the action level and restrictions on reinjection of treated or nutrient-rich waters into the aquifer system may pose major limitations with respect to biorestoration. Capture of hydrocarbon-contaminated groundwater is an engineering issue that should be readily addressed based on site-specific hydrologic considerations, whereas reinjection involves institutional issues that may be more difficult to overcome.

Reinjection of treated waters, which in some cases may be nutrient-enriched, could potentially be seen as an environmental threat in and of itself. In some situations, there could be a problem obtaining permits necessary for reinjection. Manipulation of groundwater characteristics to optimize microbial degradation may require the introduction of various growth substrates and nutrients such as nitrogen. Although these materials should be applied at the rates required for microbial metabolism, excess loadings may be needed to ensure that adequate distribution in the aquifer is provided. Because complete hydraulic isolation of the aquifer may not be possible in most situations where biorestoration is employed, movement of added materials beyond the zone of contamination may occur. These potential impacts must therefore be assessed in relation to the potential benefits derived from the biorestoration operation. In undertaking any biorestoration system, consideration must be given to the tradeoffs between injecting certain materials into the groundwater regime and remediating the gasoline components to acceptable levels.

2.4.4 Costs

Costs for biorestoration depend on specific factors such as the nature of the hydrogeology and groundwater chemistry at the site, groundwater quantity and quality at the site, the quantity of contaminants, and the required level of cleanup.

For the Biocraft site discussed in this section, the total capital cost for remediation was $926,000 (Amdurer et al., 1986). Approximately $446,000 was expended during the feasibility evaluation for the site, which was completed within 2.5 years. The operating costs for treatment of groundwater once the system was operational, exclusive of any capital expenditures, was $0.0165/gal ($225.50/day, 13,680 gpd).

Other cost estimates for biorestoration projects include:

* Ehrenfeld and Bass (1984) estimate the six-month cost for in situ biodegradation of wastes on a 1-acre site to be $1200.

* A. M. Kirby of Suntech, Inc. estimates costs of approximately $50,000 for a 5-acre site with a six-month cleanup period.

* Richard L. Raymond, Jr. of Biosystems estimates a cost of $4–$6/lb of contaminant removed (compared to $15–$20/lb for air stripping and $40/lb for carbon adsorption).

* Dr. Ralph Portier of Louisiana State University estimates a cost of $30–$50/yd³ of contaminated soil (compared to $125–$130/yd³ for hauling and disposal of the soil).

2.4.5 Summary

Biorestoration techniques provide methods whereby groundwaters contaminated with gasoline components can be effectively remediated. The technology has not been widely applied; packed air towers and carbon filtration are still the preferred technologies for groundwater treatment. Biorestoration does show promise as a cost-effective alternative. Biorestoration accomplished in situ or in bioreactors has been demonstrated to be effective in degrading hydrocarbons, although the degree of cleanup is highly dependent on specific environmental conditions affecting microbial growth. The time required for biorestoration techniques to effectively mitigate gasoline-contaminated aquifers is expected to be on the order of years, while shorter periods may be achieved using physicochemical techniques such as air stripping or carbon adsorption.

Application of biorestoration techniques should be considered for gasoline-contaminated aquifers where control of contaminant migration can be achieved hydraulically. Because these corrective action techniques involve minimal construction impacts, they are highly suited for implementation at active facilities.

One of the main advantages of biodegradation over other techniques is that the contaminants are completely destroyed, leaving the end products of aerobic degradation, carbon dioxide and water. Air stripping and activated carbon, on the other hand, are both separation techniques where the contaminant is simply transferred to a different medium. Thus, biorestoration escapes problems such as vapor-phase treatment (air stripping) or disposal/regeneration of hazardous spent activated carbon.

Biorestoration techniques can be utilized in conjunction with other physicochemical corrective actions. This may be the most beneficial aspect of these processes. In situ biorestoration techniques can be coupled with soil gas venting and groundwater extraction and treatment techniques to provide accelerated aquifer restoration. Bioreactors can also be used in conjunction with air strippers or carbon adsorption systems to provide high levels of treatment.

Because of the relatively low costs associated with biorestoration techniques as compared to other techniques, it is expected that application of these methods to gasoline-contaminated aquifers should be given serious consideration during project development and scoping.

3　SAMPLE PACKED AIR TOWER PROBLEM

This section provides a sample problem for designing a packed air tower to remove gasoline from contaminated groundwater. Because the engineering evaluations for packed air tower design are straightforward, it is possible to demonstrate a step-by-step approach to show the reader the relevant equations, the key decision points, and the eventual design parameters. Note that pilot plant testing is always recommended before full-scale implementation of any packed air tower design. With this sample problem, however, the reader can proceed with at least an initial design at the site of interest and become more aware of how some design parameters influence the design of others—with eventual cost implications for each.

3.1 SAMPLE PROBLEM

An instantaneous gasoline spill has contaminated groundwater at UST Site 1. Groundwater quality analyses show the following *average* concentrations in the gasoline plume:

Compound	Concentration (μg/L)
benzene	100
toluene	50
xylenes	200

Hydrogeologic investigations show that the groundwater can be withdrawn at a rate of 1 MGD (0.0439 m³/sec). The regulatory agency monitoring the cleanup has determined that given the volume of contaminated groundwater, a treatment flow rate of 1 MGD would be sufficient to restore the aquifer in a suitable amount of time. The agency believes the critical compound to be removed is benzene and has set a target effluent goal of 1 μg/L. The temperature of the groundwater was measured to be 15°C (288°K).

3.2   APPROACH TO PROBLEM

The overall approach to designing a packed air tower to meet the proposed cleanup goal can be broken into the following 13 steps:

1. Select compound to be stripped.

2. Select packing material.

3. Determine minimum air-water ratio.

4. Select stripping factor.

5. Determine operating air-water ratio.

6. Select design pressure drop.

7. Determine air and water loading on tower.

8. Determine tower diameter.

9. Determine mass transfer coefficient.

10. Determine height of transfer unit.

11. Determine number of transfer units.

12. Determine packing height.

13. Go to Step 4. Iterate stripping factor and/or gas pressure drops to determine design parameters that optimize cost-effectiveness.

Key design parameters are determined in each step. A summary of each design parameter, its value in the sample problem, and how it was determined is presented in Table 5.

Step 1: Select Compound to Be Stripped

In this sample problem, the decision as to which compound should be used as the basis of the packed air tower design has been simplified. The regulatory agency set a cleanup goal of 1 μg/L for benzene.

Table 5. Design Parameters for Packed Air Tower Example Problem

Variable	Description	Units	Value in Example	How Determined
a_t =	surface area/volume of packing	m^2/m^3	157	Table 1 in Step 2
a_w =	wetted area of packing per unit volume	m^2/m^3	77.9	Equation 6 in Step 9d
C_e =	effluent concentration	$\mu g/L$	1	Given
C_f =	packing factor	$1/m$	49	Table 1 in Step 2
C_i =	influent concentration	$\mu g/L$	100	Given
D =	tower diameter	m	2.49	Equation 21 in Step 8
D_a =	diffusivity in air	m^2/sec	1×10^{-6}	Equation 27 in Step 9f
d_s =	diameter of sphere with same surface area as unit of packing	m	0.062	Equation 31 in Step 9g
D_w =	diffusivity in water	m^2/sec	6.14×10^{-10}	Equation 25 in Step 9e
g =	gravitational constant	m/sec^2	9.81	Given
G =	unit air (or gas) loading rate	$m^3/m^2/sec$	0.52	Equation 19 in Step 7
H =	Henry's law constant	dimensionless	0.102	Table 7 (benzene)
HTU =	height of transfer unit	m	2.3	Equation 14 in Step 10
k_g –	air-phase mass transfer coefficient	kg-mol/m²/ atm/sec	4×10^{-5}	Equation 32 in Step 9h
k_l =	liquid-phase mass transfer coefficient	m/sec	9.63×10^{-4}	Equation 30 in Step 9g
$K_L a$ =	mass transfer coefficient	1/sec	5.67×10^{-3}	Equation 33 in Step 9i
L =	unit liquid (or water) loading rate	$m^3/m^2/sec$	0.013	Equation 20 in Step 7
M_a =	molecular weight of air	kg/kmol	28.95	Given in Step 9f
M_c =	molecular weight of compound	kg/mol	78	Given (for benzene) in Step 9f
M_w =	molecular weight of water	kg/kmol	18.2	Given in Step 9e
N_F =	Froude number	dimensionless	0.0027	Equation 23 in Step 9b
N_R =	Reynolds number	dimensionless	72.7	Equation 22 in Step 9a
NTU =	number of transfer units	–	5.7	Equation 13 in Step 10
N_W =	Weber number	dimensionless	0.146	Equation 24 in Step 9c
P_t =	operating pressure of tower	atm	1	Given
Q =	flow rate	m^3/sec	0.0439	Given
R =	stripping factor	dimensionless	4 to 4.08	Assume R=4 and then iterate
SF_K =	safety factor for $K_L a$	–	1.2	Given in Step 9i
T =	temperature of water	°K	288	Given

Table 5,   continued

Variable		Description	Units	Value in Example	How Determined
T_c	=	surface tension of packing	kg/sec^2	0.033	Table 6 in Step 2
T_w	=	surface tension of water	kg/sec^2	7.39×10^{-2}	Given in Step 9c
V_c	=	compound's molal volume at boiling point	m^3/kmol	0.096	Equation 26 in Step 9e
Z_t	=	packing height	m	13.2	Step 12
ρ_G	=	density of air	kg/m^3 at °C	1.23	Given
ρ_w	=	density of water	kg/m^3 at °C	999.1	Table 8 using 15°C (288°K)
μ_G	=	viscosity of air	kg/m/sec at °C	1.74×10^{-5}	Given in Step 9h
μ_w	=	viscosity of water	kg/m/sec at °C	1.139×10^{-3}	Table 8 using 15°C (288°K)
ψ	=	association factor	(1/m)	49	Table 1 in Step 2

Each step is discussed individually below.

Often this decision is not straightforward. Many times, cleanup goals are set for several compounds, or for groups of compounds such as BTEX (benzene, toluene, ethylbenzene, and xylene) or total hydrocarbons. If the gasoline release had occurred several years ago, it is likely that the more mobile hydrocarbons like BTEX would have volatilized or biodegraded to nondetectable levels, in which case other, more persistent compounds, especially those with molecular weights higher than BTEX, might be used for setting cleanup goals.

Step 2: Select Packing Material

Removal efficiencies in packed air towers are directly related to the type of packing material used in the tower. There are numerous packing types available. (See Table 1.) Consult manufacturers' information for additional details on particular types of packing. For the sample problem, select 2-in. plastic (Jaeger) tri-packs as the packing material. These have the following characteristics:

- packing size: 2 in. (0.050 m)

- critical surface tension of packed material (from Table 6): 0.033 kg/sec^2

- packing factor (from Table 1): C_f = 16/ft (49/m)

- surface area of packing (from Table 1): 48 ft^2/ft^3 (157 m^2/m^3)

Table 6. Critical Surface Tension of Packing Materials

Material	T_c (kg/sec^2)
Carbon	0.056
Ceramic	0.061
Glass	0.073
Paraffin	0.020
Plastic	0.033
Polyvinylchloride	0.040
Steel	0.075

Step 3: Determine Minimum Air-Water Ratio

The minimum air-water ratio is determined from Equation 8:

$$\left(\frac{G}{L}\right)_{min} = \left(\frac{C_i - C_e}{C_i}\right) \times \left(\frac{1}{H}\right) \tag{8}$$

where

$$\left(\frac{G}{L}\right)_{min} = \text{minimum air-water ratio}$$
$$G = \text{air flow rate (m}^3\text{/m}^2\text{/sec)}$$
$$L = \text{water flow rate (m}^3\text{/m}^2\text{/sec)}$$
$$H = \text{Henry's constant (dimensionless)}$$
$$C_i, C_e = \text{concentrations of influent and effluent (µg/L)}$$

$$\left(\frac{G}{L}\right)_{min} = \frac{(100 - 1)}{(100)(0.102)}$$

$$\left(\frac{G}{L}\right)_{min} = 9.7$$

Step 4: Select Stripping Factor

The stripping factor, R, represents the ratio of the operating air-water ratio to the theoretical minimum air-water ratio. R is one of the key parameters used in balancing cost and performance. As discussed in the previous chapter, optimum R values for removing petroleum hydrocarbons typically fall between 3.0 and 5.0. For this example, an initial value of 4.0 was assumed for R. At the end of the design, different values of R will be iterated to determine the most cost-effective value.

Table 7. Henry's Law Constants for Selected Organics

Contaminant	Henry's Law Constant at 20°C (Atm)
Trichloroethylene	324
Tetrachloroethylene	565
Carbon tetrachloride	1,290
1,1,1-Trichloroethane	400
Vinyl chloride	355,000
Benzene	133
Dichlorobenzene(s)	190
1,1,2-Trichloroethane	43
Toxaphene	3,500
Chloromethane	480
1,2,4-Trimethylbenzene	345
Toluene	335
Chloroform	170
Bromodichloromethane	118
Dibromochloromethane	47
Bromoform	35
Pentachlorophenol	0.12
Dieldrin	0.0094
Methylene chloride	132
Chlorobenzene	140
Trichlorobenzene	160
1,1-Dichloroethylene	346
trans-1,2-Dichloroethylene	155

Step 5: Determine Operating Air-Water Ratio

The operating air-water ratio is determined using Equation 9:

$$R = \frac{\left(\dfrac{G}{L}\right)_{\text{actual (or operating)}}}{\left(\dfrac{G}{L}\right)_{\text{min}}} \qquad (9)$$

$$\left(\frac{G}{L}\right)_{\text{actual}} = R\left(\frac{G}{L}\right)_{\text{min}}$$

$$= (4)(9.7)$$

$$\left(\frac{G}{L}\right)_{\text{actual}} = 38.8$$

To simplify calculations, round off the air-water ratio to 40:1. Re-calculate R.

$$R = \frac{40}{9.7}$$

$$R = 4.12$$

Step 6: Select Design Pressure Drop

Generalized pressure drop curves from the air moving across the packing material are presented in Figures 8 and 9 in English and metric units, respectively. Typically, using lower pressure drops results in lower horsepower requirements for the air blowers. At the same time, however, the magnitude of the pressure drop affects the sizing of the tower diameter—and consequently, construction costs. Lower pressure drops result in larger tower diameters for a given packing size. Pressure drops in packed air towers should be kept between 50 and 200 newtons $(N)/m^2/m$. A pressure drop of 100 $N/m^2/m$ can typically be used for designing a packed air tower to remove petroleum hydrocarbons. Assume a pressure drop of 100 $N/m^2/m$.

Step 7: Determine Air and Water Loading on Tower

Determine the x-axis value for the generalized pressure drop curves using Equations 17 and 18:

$$\frac{\text{air-water}}{\text{ratio}} \cdot \left(\frac{\rho_G}{\rho_w} \right) = \frac{G^L}{L^L} \qquad (17)$$

where

ρ_G = density of air (kg/m^3)
ρ_w = density of water (kg/m^3)
G^L = volumetric air flow rate (m^3/s)
L^L = volumetric water flow rate (m^3/s)

Using Table 8 and a water temperature of 15°C (288°K), determine the water density to be 999.1 kg/m^3. The air density is 1.23 kg/m^3 at 15°C.

$$40 \cdot (1.23/999.1) = \frac{G^L}{L^L}$$

$$0.049 = \frac{G^L}{L^L}$$

Table 8. Density and Viscosity of Water

Temperature (°C)	Density (kg/m³)	Dynamic Viscosity (kg/m/sec)	Surface Tension (kg/sec²)
0	999.8	1.787×10^{-3}	7.56×10^{-2}
5	1000.0	1.519×10^{-3}	7.49×10^{-2}
10	999.7	1.307×10^{-3}	7.42×10^{-2}
15	999.1	1.139×10^{-3}	7.35×10^{-2}
20	998.2	1.002×10^{-3}	7.28×10^{-2}
25	997.0	0.890×10^{-3}	7.20×10^{-2}

Source: Handbook of Chemistry and Physics, 59th edition (Boca Raton, FL: CRC Press, 1978–1979).

Invert $\dfrac{G^L}{L^L}$ and put into Equation 18:

$$\text{x-axis} = \left(\frac{L^L}{G^L}\right)\left(\frac{\rho_G}{\rho_w - \rho_G}\right)^{1/2} \tag{18}$$

$$= (1/0.049)(1.23/[999.1 - 1.23])^{1/2}$$
$$\text{x-axis} = 0.72$$

Enter Figure 9, with a value of 0.72, reflect off the 100 N/m²/m curve, and obtain a value for the collective y-axis term of 0.0083.

The air (or gas) loading rate can now be determined from Equation 19:

$$\text{y-axis} = \frac{(G^2\, C_f\, \mu_w^{0.1}\, J)}{(\rho_G\, (\rho_w - \rho_G)\, g_c)} \tag{19}$$

where

G = air loading rate (m³/m²/sec), calculated in kg/m²/sec and then converted using air density
C_f = packing factor (1/m)
μ_w = dynamic viscosity of water (kg/m/sec)
J = coefficient that accounts for types of phase and is equal to 1 for water/air systems
ρ_G = density of air (kg/m³)
ρ_w = density of water (kg/m³)
g_c = conversion factor equal to 1 for metric units (dimensionless)

Solve for G by taking C_f from Table 1 and ρ_w and μ_w from Table 8 for 15°C:

$$0.0083 = \frac{G^2 \, (49)(1.139 \times 10^{-3})^{0.1}(1)}{(1.234)(999.1 - 1.234)(1)}$$

$$G^2 = 0.41$$

$$G = 0.64 \text{ kg/m}^2/\text{sec or } \frac{0.64}{1.234} = 0.52 \text{ m}^3/\text{m}^2/\text{sec}$$

The unit water (liquid) loading rate can now be determined from the air-water ratio:

$$\frac{\text{air-water}}{\text{ratio}} = \frac{G}{L} \qquad (20)$$

where

G = unit air loading rate (m^3/m^2/sec)
L = unit liquid loading rate (m^3/m^2/sec)

$$40 = \frac{0.52}{L} \text{ or } L = 0.013 \text{ m}^3/\text{m}^2/\text{sec}$$

Because 1 m^3 = 1000 kg, L also equals 13 kg/m^2/sec.

Step 8: Determine Tower Diameter

Next, the tower diameter is determined using Equation 21:

$$D = \left[\frac{4Q}{\pi L} \right]^{0.5} \qquad (21)$$

where

D = tower diameter (m)
Q = flow rate (m^3/sec)
L = unit liquid loading rate (m^3/m^2/sec)

$$D = \left[\frac{(4) \, (0.0439)}{\pi \, (0.013)} \right]^{0.5}$$

$$D = 2.07 \text{ m}$$

Applying a safety factor of 1.2, D = 2.49 m.

Step 9: Determine Mass Transfer Coefficient

Determining the mass transfer coefficient, $K_L a$, for a contaminant in water and in air is a complex undertaking. If a mass transfer coefficient has already been determined at your site from pilot plant studies or other sources, you can proceed to Step 10. If a mass transfer coefficient is not available for conditions at your site, one can be determined as follows.

Step 9a.   Determine Reynolds number.

Step 9b.   Determine Froude number.

Step 9c.   Determine Weber number.

Step 9d.   Determine actual wetted area of packing.

Step 9e.   Determine contaminant's diffusivity in water.

Step 9f.   Determine contaminant's diffusivity in air.

Step 9g.   Determine diffusion coefficient in water.

Step 9h.   Determine diffusion coefficient in air.

Step 9i.   Determine mass transfer coefficient.

Each step is discussed below.

Step 9a: Determine Reynolds Number

Reynolds number, N_R, can be determined from Equation 22. (Note units for L.)

$$N_R = \frac{L}{\mu_w \, a_t} \tag{22}$$

where

$$L = \text{unit liquid loading rate (kg/m}^2\text{/sec)}$$
$$\mu_w = \text{dynamic viscosity (kg/m/sec)}$$
$$a_t = \text{surface area per volume of packing (m}^2\text{/m}^3\text{)}$$

$$N_R = (13)/(1.139 \times 10^{-3})(157)$$
$$= 72.7 \text{ (dimensionless)}$$

Step 9b: Determine Froude Number

Froude number, N_F, can be determined from Equation 23. (Note units for L.)

$$N_F = \frac{L^2 \, a_t}{\rho_w^2 \, g} \tag{23}$$

where

L = unit liquid loading rate (kg/m²/sec)
a_t = surface area per volume of packing (m²/m³)
ρ_w = density of water (kg/m³)
g = gravity constant = 9.81 m/sec²

$$N_F = (13)^2(157)/(999.1)^2(9.81)$$
$$= 0.0027 \text{ (dimensionless)}$$

Step 9c: Determine Weber Number

Weber number, N_W, can be determined from Equation 24. (Note units for L.)

$$N_W = \frac{L^2}{\rho_w \, T_w \, a_t} \tag{24}$$

where

L = unit liquid loading rate (kg/m²/sec)
ρ_w = density of water (kg/m³)
T_w = surface tension of water (kg/sec²)
a_t = surface area per volume of packing (m²/m³)

$$N_W = (13)^2/(999.1)(7.39 \times 10^{-2})(157)$$
$$= 0.0145$$

Step 9d: Determine Actual Wetted Area of Packing

The actual wetted area of the packing material can be determined from Equation 6:

$$\frac{a_w}{a_t} = 1 - \exp\left(- 1.45\left[\frac{T_c}{T_w}\right]^{0.75} \left[N_R\right]^{0.1} \left[N_F\right]^{-0.05} \left[N_W\right]^{0.2}\right) \tag{6}$$

where

a_w = wetted area (m²/m³)
a_t = surface area per volume of packing (m²/m³)

$$T_c = \text{surface tension of packing (kg/sec}^2)$$
$$T_w = \text{surface tension of water (kg/sec}^2)$$
$$N_R, N_F, N_W = \text{Reynolds, Froude, Weber numbers}$$
$$\text{(dimensionless)}$$

$$\frac{a_w}{157} = 1 - \exp\left(-1.45\left[\frac{0.033}{0.0739}\right]^{0.75}\left[72.7\right]^{0.1}\left[0.0027\right]^{-0.05}\left[0.0145\right]^{0.2}\right)$$

$$a_w = 157\,(1 - \exp\,[-0.7])$$

$$a_w = 77.9 \text{ m}^2/\text{m}^3$$

The wetted area, a_w, is used to determine the liquid-phase mass transfer coefficient in Step 9g.

Step 9e: Determine Contaminant's Diffusivity in Water

The diffusivity of benzene in water can be determined from:

$$D_w = \frac{(117.3 \times 10^{-18})(\psi\,M_w)^{0.5}\,T}{\mu_w^{1.14}\,V_c^{0.5}} \tag{25}$$

where

$$D_w = \text{diffusivity of compound in water (m}^2/\text{sec)}$$
$$\psi = \text{an association factor between liquid and air. For air/}$$
$$\text{water systems such as packed air towers, the associa-}$$
$$\text{tion factor is 2.26 (dimensionless).}$$
$$M_w = \text{molecular weight of water (18.2 kg/kmol)}$$
$$T = \text{water temperature (}^\circ\text{K)}$$
$$\mu_w = \text{viscosity of water at 15}^\circ\text{C (kg/m/sec)}$$
$$V_c = \text{the compound's molal volume at the normal boiling}$$
$$\text{point (m}^3/\text{kmol)}$$

V_c can be determined for benzene by taking the atomic volumes for carbon and hydrogen from Table 9 and the number of carbon and hydrogen atoms in benzene from Table 10 and using:

$$V_c = \frac{\left[\left(\begin{array}{c}\text{atomic volume}\\\text{of carbon}\end{array}\right)\left(\begin{array}{c}\text{No. of}\\\text{carbon}\\\text{atoms}\end{array}\right) + \left(\begin{array}{c}\text{atomic}\\\text{volume}\\\text{of}\\\text{hydrogen}\end{array}\right)\left(\begin{array}{c}\text{No. of}\\\text{hydrogen}\\\text{atoms}\end{array}\right) - 15\right]}{1000} \tag{26}$$

Table 9. Molal Volume of Typical Atomic and Molecular Components

Atomic Volume $(m^3/1{,}000\ atoms \times 10^3)$		Molecular Volume $(m^3/kmol \times 10^3)$	
Carbon	14.8	H_2	14.3
Hydrogen	3.7	O_2	25.6
Chlorine	24.6	N_2	31.2
Bromine	27.0	Air	29.9
Iodine	37.0	CO	30.7
Sulphur	25.6	CO_2	34.0
Nitrogen	15.6	SO_2	44.8
In primary amines	10.5	NO	23.6
In secondary amines	12.0	N_2O	36.4
Oxygen	7.4	NH_3	25.8
In methyl esters	9.1	H_2O	18.9
In higher esters	11.0	H_2O	32.9
In acids	12.0	COS	51.5
In methyl ethers	9.9	Cl_2	48.4
In higher ethers	11.0	Br_2	53.2

If benzene rings are present in the compound, subtract 15 for each ring. If naphthalene rings are present, subtract 30 for each double ring.

Source: Treybal, 1980.

$$= \frac{[(14.8)(6) + (3.7)(6) - 15]}{1000}$$

$$= 0.096\ m^3/kmol$$

Note that 15 is subtracted for benzene rings only. If naphthalene rings are present, subtract 30 for each double ring.

The diffusivity, D_w, can now be determined.

$$D_w = (117.3 \times 10^{-18})\ [(2.26)(18.2)]^{\,0.5}\ (288)/(1.139 \times 10^{-3})(0.096)^{\,0.5}$$

$$= 6.14 \times 10^{-10}\ m^2/sec$$

Step 9f: Determine Contaminant's Diffusivity in Air

The diffusivity of benzene in air, D , can be determined from:

$$D_a = \frac{10^{-4}(1.084 - 0.249)\ \sqrt{M_c^{-1} + M_a^{-1}}\,(T^{1.5})\sqrt{M_c^{-1} + M_a^{-1}}}{P_t\ \tau^2\ f(KT/\in)} \quad (27)$$

where

$$D_a = \text{diffusivity of compound in air } (m^2/sec)$$

M_c = molecular weight of compound (kg/kmol). For benzene, M = 78 kg/kmol.

M_a = molecular weight of air = 28.95 kg/kmol

T = temperature of water (°K)

P_t = absolute pressure of packed tower system in N/m^2. Pressure usually is 1 atm = 101,325 N/m^2.

τ = molecular separation at collision as determined from chemical characteristics

K = Foltzmann's constant = 1.3805×10^{-23} J/K

$f(KT/\in)$ = collision function (from Figure 35)

$\in$ = energy of molecular attraction

Before solving Equation 27, it is necessary to undertake some preliminary calculations:

$$\frac{\in}{K} = 1.21\ T_b \qquad\qquad (28)$$

where

T_b = normal boiling point of compound (°K)

For benzene, Table 10 shows a boiling point of 353°K. Thus:

$KT/\in$ = $(1/1.2\ T_b)T$ or
$KT/\in$ = $(1/(1.2)(353)(288)$
$KT/\in$ = 0.67

Using the collision function nomograph in Figure 35, enter the x-axis with $KT/\in$ = 0.67 and determine $f(KT/\in)$ from the y-axis. $f(KT/\in)$ = 0.91.

The molecular separation, τ, can be determined next from:

$$\tau = \frac{\tau_A + \tau_c}{2} \qquad\qquad (29)$$

where

τ = molecular separation at collision

τ_A = molecular radius of air = 0.3711

τ_c = molecular radius of compound to be stripped = $1.18\ V_c^{0.33}$

V_c = molal volume of compound at normal boiling point (m^3/kmol)

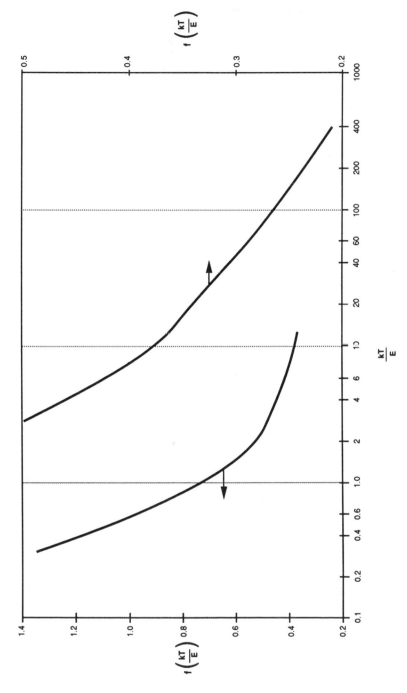

Figure 35. Collision function for diffusion.

Table 10. Properties of Certain Compounds

Contaminant	Molecular Weight (g/m)	Boiling Point (°C)	Boiling Point (°K)	Molal Volume (m³/kmol)
Benzene	78	80.1	353	0.0969
Carbon tetrachloride	154	77	350	0.1132
Chlordane	406	175	448	0.3760
Dichlorobenzene	147	180	453	0.1378
1,2-Dicholoroethane	99	83	356	0.0936
1,1-Dicholoroethylene	97	36	309	0.0862
trans-1,2-dichloroethylene	97	47	320	0.0862
Methylene chloride	85	40	313	0.0714
Polychlorinated biphenyls	267	345	618	0.2473
Tetrachloroethylene	166	121	394	0.1280
1,1,1-Trichloroethane	133	74	347	0.1145
Trichloroethylene	131	87	360	0.1071
Vinyl chloride	62	-13.4	260	0.0653
Xylene	106	144	417	0.1404
Water	18	100	373	0.0148

Substituting $1.18 \ V_c^{0.33}$ for τ_c in Equation 29 gives:

$$\tau = (\tau_A + 1.18 \ V_c^{0.33})/2$$
$$= (0.3711 + 1.18 \ (0.96)^{0.33})/2$$
$$= 0.456$$

With these variables determined, Equation 27 for determining a compound's diffusivity in air can now be solved:

$$D_a = \frac{10^{-4} \ (1.084 - 0.249) \ \sqrt{^1/_{78} + \ ^1/_{28.95}} \ (288)^{1.5} \ \sqrt{^1/_{78} + \ ^1/_{28.95}}}{(101,325)(0.456)^2 \ (0.91)}$$

$$= 1 \times 10^{-6} \ \text{m}^2/\text{sec}$$

Step 9g: Determine Diffusion Coefficient in Water

A compound's diffusion coefficient in water, k_l, can be determined from:

$$k_l = 0.0051 \left[\frac{L \ \rho_w}{\mu_w \ a_w} \ \frac{\mu_w}{\rho_w \ D_w} \right]^{-0.5} \left[a_t \ d_s \right]^{0.4} \left[\frac{\rho_w}{\mu_w \ g} \right]^{-0.33} \tag{30}$$

where

L = unit liquid loading rate (m³/m²/sec)
ρ_w = density of water (kg/m³)

μ_w = dynamic viscosity of water (kg/m/sec)
a_w = wetted area of packing (m²/m³)
D_w = diffusivity in water (m²/sec)
a_t = total surface area of packing (m²/m³)
g = gravitational constant (m/sec²)
d_s = diameter of a sphere with the same surface area as a piece of packing material (m)

$$d_s = 2 \sqrt{\frac{0.012}{4\pi}} \text{ or } 0.062 \text{ m}$$

$$k_l = (0.0051) \left[\frac{(0.013)(999.1)}{(1.139 \times 10^{-3})(77.9)} \right] \left[\frac{1.139 \times 10^{-3}}{(999.1)(6.14 \times 10^{-10})} \right]^{-0.5} \times$$

$$\left[157(0.062) \right]^{0.4} \left[\frac{999.1}{(1.139 \times 10^{-3})(9.81)} \right]^{-0.33} \qquad (31)$$

$= (0.0051)(146.38)(0.0232)(2.48)(0.02236)$

$= 9.63 \times 10^{-4}$ m/sec

9h: Determine Diffusion Coefficient in Air

The air phase mass transfer coefficient can be determined from:

$$k_g = 5.23 \left[\frac{G \, \rho_G}{\mu_G \, a_t} \right]^{0.7} \left[\frac{\mu_G}{\rho_G \, D_a} \right]^{0.33} \left[a_t \, d_s \right]^{-2} \left[\frac{R' \, T}{a_t \, D_a} \right]^{-1} \qquad (32)$$

where

k_g = air phase mass transfer coefficient (kg-mol/m²-atm-sec)
G = unit air (or gas) loading rate (m³/m²/sec)
ρ_G = density of air at °C (kg/m³)
μ_G = viscosity of air = 1.74×10^{-5} kg/m/sec at 15°C
a_t = total surface area of packing
D_a = diffusivity of compound in air (m²/sec)
d_s = diameter of a sphere with same surface area as one unit of packing material (m)
R' = Universal Gas Constant = 0.0821 atm-m³-km⁻¹-°K⁻¹
T = absolute temperature of water (°K)

$$k_g = 5.23 \left[\frac{(0.52)(1.23)}{(1.74 \times 10^{-5})(157)}\right]^{0.7} \left[\frac{(1.74 \times 10^{-5})}{(1.23)(1 \times 10^{-6})}\right]^{0.33} \times$$

$$\left[157(0.062)\right]^{-2} \left[\frac{(0.0821)(288)}{(157)(1 \times 10^{-6})}\right]^{-1}$$

$$= 5.23 \, (45.56)(2.397)(0.0106)(0.0000066)$$

$$= 4 \times 10^{-5} \text{ kg-mol/m}^2\text{-atm-sec}$$

Step 9i: Determine Mass Transfer Coefficient

The mass transfer coefficient, $k_L a$, can be determined from the following:

$$k_L a = \frac{a_w \left[\dfrac{1}{k_l} + \dfrac{\rho_G}{(28.95)(kg\,H)}\right]^{-1}}{SF_k} \qquad (33)$$

where

$k_L a$ = mass transfer coefficient $\left(\dfrac{1}{sec}\right)$

a_w = wetted area of packing per unit volume (m^2/m^3)

k_l = liquid phase mass transfer coefficient (m/sec)

k_g = air phase mass transfer coefficient (kg-mol/m^2-atm-sec)

ρ_G = density of air (kg/m^3)

H = Henry's constant for a particular compound at a particular temperature (dimensionless)

SF_k = safety factor (dimensionless)

Safety factors between 1.2 and 2 are typically used. For this sample problem SF_k was set equal to 1.2 (also used for the tower diameter in Step 8).

$$k_L a = \frac{77.9 \left[\dfrac{1}{(9.63 \times 10^{-4})}\right] + \left[\dfrac{1.23}{(28.95)(4 \times 10^{-5})(0.102)}\right]^{-1}}{1.2}$$

$$= 5.67 \times 10^{-3}/\text{sec}$$

Step 10: Determine Height of Transfer Unit

The height of a transfer unit, HTU (in m), is inversely related to the mass transfer coefficient:

$$\text{HTU} = \frac{L}{k_L a} \qquad (14)$$

where

$$L = \text{unit liquid loading rate } (m^3/m^2/sec)$$

$$k_L a = \text{mass transfer coefficient } \left(\frac{1}{sec}\right)$$

$$\text{HTU} = 0.013/5.67 \times 10^{-3}$$
$$= 2.3 \text{ m}$$

Step 11: Determine Number of Transfer Units

The number of transfer units, NTU, can be determined from:

$$\text{NTU} = \left(\frac{R}{R-1}\right) \ln \left[\frac{((C_i/C_e)(R-1)+1)}{R}\right] \qquad (13)$$

where

$$R = \text{stripping factor}$$
$$C_i = \text{influent concentration } (\mu g/L)$$
$$C_e = \text{effluent concentration } (\mu g/L)$$

Thus:

$$\text{NTU} = \left(\frac{4.08}{4.08-1}\right) \ln \left[\frac{((100/1)(4.08-1)+1)}{4.08}\right]$$
$$= 5.73 \text{ units}$$

Step 12: Determine Packing Height

The packing height (Z_t) needed to decrease benzene concentrations from 100 μg/L to 1 μg/L can be determined by:

$$Z_t = (\text{NTU})(\text{HTU}) \qquad (34)$$

Thus:

$$Z_t = (5.73)(2.3)$$
$$= 13.2$$

The first iteration of the sample design is completed. To reduce benzene concentrations from 100 μg/L to 1 μg/L in a 1-MGD treatment system requires:

tower diameter = 2.49 m
packing height = 13.2 m
air-water ratio = 40:1

Water would be in contact with the packing material for 1015 sec or 17 min (13.2 m/0.013 m^3/m^2/sec = 1015 seconds).

There are other combinations of air-water ratios and gas pressure drops that will provide the same removal efficiency. At higher air-water ratios, pressure drops will be significantly higher. Higher pressure drops require larger blowers, resulting in an overall increase in energy costs. There are trade-offs that must be made in designing the optimal system. The final step in the design is to revisit Step 4 and iterate different stripping factors and gas pressure drops, determining the costs of each design. To simplify the sample problem, only one design will be carried into the cost determinations.

3.3   CAPITAL COSTS OF AIR STRIPPING

Packed Tower Cost

The capital costs of the packed air tower designed in the previous section can now be determined using cost curves developed by Camp Dresser & McKee (1987) and shown in Figures 11 through 14. Reiterating the key design features:

- flow rate = 0.0439 m^3/sec (1 MGD)
- diameter of tower = 2.49 m (8.2 ft)
- depth of packing = 13.2 m (43.3 ft)
- air-water ratio = 40:1

The surface area of the packed air tower is needed for Figure 11, and can be determined by πr^2, or $\pi(8.2/2)^2$ = 52.8 ft^2.

Entering Figure 11 with a surface area of 52.8 ft^2 and interpolating a packing depth of 43.3 ft gives a capital cost of $210,000.

Clearwell Cost

The cost of a clearwell to treat 1 MGD of water can be determined directly from Figure 12 to be $45,000.

Pump Costs

The pump costs associated with treating 1 MGD can be determined directly from Figure 13 to be $26,000.

Blower Costs

Using an air-water ratio of 40:1 and knowing that 1 MGD = 1 CFM, the air flow rate can be determined from:

$$AF = (AW)(Q) \tag{35}$$

where
AF = air flow rate (CFM)
AW = air-water ratio

Q = water flow rate (CFM)
AF = 40×93
= 3720 CFM

Entering Figure 14 with an air flow rate of 3720 CFM and a head loss of 9 in., the cost of one air blower is about $10,000. The total capital cost for the system is:

Tower	$210,000
Clearwell	45,000
Pump	26,000
Blower	10,000
TOTAL CAPITAL COST:	$291,000

If air pollution controls are required because the emissions from the tower exceed regulatory standards, then these costs must be added.

3.4 OPERATION AND MAINTENANCE COSTS OF AIR STRIPPING

The O&M costs of the packed air tower can usually be determined from Figures 15 and 16 (assuming a power cost of 9.6¢ per kilowatt-hour).

Pump Operation

Figure 15 is based on the relationship:

$$\text{pump cost} = 0.006 + 0.0006Z_t \tag{36}$$

For Z_t = 43.3 ft, the associated costs to operate a pump would be $(0.006) + (0.006)(43.3)$ = 3.2¢/1000 gal treated.

Blower Operation

The related costs to operate the air blower for an air-water ratio of 40:1 and 9 in. W.C. would be 3.5¢/1000 gal treated.

Adding in typical labor costs of 2¢/1000 gal, the O&M costs total 8.7¢ per 1000 gal. Assuming the 1-MGD treatment facility is operating at design capacity for one full year, the annual operating cost would be:

$$\frac{O \& M}{Cost} = \left(\frac{8.7¢}{1000 \text{ gal}} \right) (365 \text{ days}) \left(\frac{1 \times 10^6 \text{ gal}}{\text{day}} \right)$$

$$= \$31,755 \text{ per year}$$

3.5 CONCLUSION

The packed air tower was sized to achieve 99% removal of benzene. However, other volatile compounds would also be removed. At the beginning of the sample problem, toluene and xylene were also cited as contaminants found in the gasoline plume. The packed air tower designed in the sample problem would remove both of these volatile compounds. The Henry's law constants for toluene (0.27) and xylene (0.29) are both greater than the Henry's law constant for benzene (0.23) so that even greater removal efficiencies could theoretically be expected for these two compounds with the present design.

After pilot testing and full-scale implementation, the performance of the packed air tower would be monitored to evaluate actual removal efficiencies compared to the idealized removal efficiency of the sample problem. Benzene effluent concentrations may be in the low ppb, but may still exceed the cleanup goal of 1 ppb. In this case, the decision may be to consider the effluent concentrations as "close enough" given the limitations of the treatment technology. If conditions at the site warrant that the cleanup goal of 1 ppb must be met, then additional polishing with GAC is needed. Designing carbon systems is not as straightforward as designing packed air towers. Removal efficiencies for GAC systems depend totally on the unique chemical characteristics of the influent stream, an evaluation that cannot be "cookbooked." Pilot testing is essential for determining carbon needs and costs.

APPENDIX A   FEDERAL REGULATIONS

On September 23, 1988, the United States Environmental Protection Agency (EPA) issued comprehensive regulations affecting owners and operators of underground storage tank systems throughout the United States. The regulations apply to tanks that store petroleum or hazardous substances (definitions of these terms and others used in the regulations are given later in this appendix), with some exceptions. The regulations are summarized in Figure 36 and discussed in greater detail below.

The major objective of the EPA regulations is to reduce the risks to human health and the environment from underground storage tank releases. To reach this goal, the regulations incorporate three broad strategies: to identify and then to correct faulty or leaking tanks; to reduce the incidence of future releases by, for example, mandating minimum operating and performance standards; and to minimize the hazards from releases that do occur by mandating a standard release investigation, response, and corrective action procedure.

An integral step in attaining the goals of these regulations is the identification of existing tank systems that have leaked in the past, are leaking presently, or may leak in the near future. Toward this end, the UST regulations require that all tank systems meet the specified performance standards by December 22, 1998, or be permanently closed. This stipulation will force most tank owner/operators either to upgrade or replace existing tanks by that date. During the process of upgrading, replacing, or closing of the tanks, faulty or leaking systems will be identified.

Indeed, if the past is prologue to the future, thousands of faulty and leaking tanks will be found. In its *Federal Register* proposal (April 17, 1988), EPA cited three studies of "non-tight" UST systems. Of these three studies, two (one an EPA-sponsored report, the other sponsored by Chevron) reported results from tank tightness tests while one discussed the results of Suffolk County, New York's UST program. The percentage of tanks failing tightness tests ranged from 10% in the Chevron study to 35% in the EPA study. Suffolk County, whose UST program predates the federal program, reported 26% of the tanks tested had failed. A more recent study by Suffolk County showed that 29% of 500 tanks that were removed contained perforations; about 60% of these showed evidence of leakage.

Such results suggest a large number of releases will be confirmed in the years to come. Removal, replacement, or upgrading of faulty and leaking tanks over the next several years and the remediation of the associated releases should reduce future environmental and health risks.

A second strategy for lowering risks is to reduce the incidence of leaks and other releases (like spills and overfills) in the future. The UST regulations facilitate this by mandating standards for the design, construction, and installation of new tanks; standards for the operation of USTs, including recordkeeping; and requirements for release prevention and detection. By stipulating minimum acceptable standards in these areas, EPA hopes that in the future the leak incidence rate will decline.

The third area where future risks may be minimized is streamlining and standardizing the ways in which owners and operators respond to suspected and confirmed releases from USTs. The regulations contain two subparts, one dealing with release reporting, investigation, and confirmation, and one with release response and corrective actions for those reported releases. Prior to enactment of these regulations, there were no standard nationwide procedures to respond to suspected or confirmed releases. Now, owners and operators of tanks suspected of leaks must report suspected releases to the implementing agency (state or local agency with responsibility over the tanks) within 24 hours and take action within one week to confirm the suspected release or begin corrective action at the site. In addition, the regulations both specify action that must be initiated to mitigate the release and set a schedule for that response. Through these actions, damage from releases should be minimized.

The aggressive implementation of these regulations in future years will likely result in the discovery of a large number of leaking USTs. While not every release will require remediation, a response will be necessary (1) where water supplies are threatened, (2) where the release poses a danger to human health and the environment, or (3) where a cleanup is politically desirable. The measures discussed elsewhere in this book will likely be used extensively to restore groundwater.

OVERVIEW OF REGULATIONS

Below is a summary of the UST regulations. The EPA regulations serve as the basis for underground storage tank regulations throughout the country, and in many cases will be those in effect in your area. However, many state and local governments have passed their own versions of the UST regulations. In some cases, these state and local regulations may differ from the federal regulations, and may in fact be more stringent. Readers are urged to read and understand the regulations in effect for their locality.

Who Is Regulated

Generally, all owners and operators of underground tanks that store "regulated" substances must comply with the EPA regulations. "Regulated" substances include petroleum products and any substance defined in section 101(14) of the Comprehensive Environmental Response, Compensation, and Liability Act of 1980 (CERCLA). Not included are any substances regulated under Subtitle C of the Solid Waste Disposal Act (SWDA) as hazardous waste.

Several classes and types of tank systems are not affected by these regulations. There are two types of exclusions: statutory exclusions (tank systems that Congress specifically exempted from these regulations) and regulatory exclusions (tanks that EPA exempted after determining that they do not pose a danger to human health or the environment or that are already covered by other regulations).

Statutory Exclusions

1. farm or residential tanks of 1100 gallons capacity or smaller used for storing motor fuel for noncommercial purposes

2. tanks that store heating oil for use on the premises where they are stored

3. septic tank systems

4. pipeline facilities regulated under specified federal or state laws

5. any surface impoundments, pits, ponds, or lagoons

6. storm water or wastewater collection systems

7. flow-through process tanks

8. liquid traps or associated gathering lines directly related to oil or gas production and gathering operations

9. tanks in underground areas

EPA Regulatory Exclusions

1. tanks that store (a) hazardous wastes listed or identified under Subtitle C of the Solid Waste Disposal Act, or (b) a mixture of such hazardous waste and regulated substances

2. wastewater treatment tank systems regulated under Section 402 or 307(b) of the Clean Water Act

3. equipment or machinery containing regulated substances for operational purposes, such as hydraulic lift tanks.

4. USTs whose capacity is 110 gallons or less

5. USTs containing *de minimis* concentrations of regulated substances

6. any emergency spill or overflow containment system that is expeditiously emptied after use

Definitions of the terms used above and of all the important terms used in the regulations are provided in the glossary (see p. 122).

The above exclusions excuse many tanks from potential regulation. For example, home heating oil tanks number in the millions. Pieces of equipment and machinery using regulated substances for operational purposes (mainly hydraulic lifts, used in elevators and service stations throughout the United States) number in the hundreds of thousands. The three *de minimis* exclusions—for small-capacity tanks, tanks that contain dilute concentrations, and tanks that hold substances only temporarily—also remove millions of tanks from the ranks of the potentially regulated. Besides serving other purposes, statutory and regulatory exclusions allow the various implementing agencies to concentrate their efforts on tanks at service stations and other commercial and industrial operations, the tanks that presumably are the focus of these regulations. Nevertheless, owners and operators of excluded tanks may still be liable for releases, and thus proper procedures should still be followed for tanks that do not fall under the scope of the UST regulations.

The UST regulations are organized into six sections, each of which is summarized below.

Subpart B: Design, Construction, Installation, and Notification Standards [280.20]

The UST regulations include requirements for the design, construction, and installation of all tanks. These performance standards separately address new tanks (installation commenced after December 22, 1988) and upgraded (existing) tanks, and aim "to prevent releases due to structural failure, corrosion, or spills and overfills as long as the UST system is used to store regulated substances."

For new tanks, four construction options are specified that meet the requirements. The tank may be constructed of fiberglass-reinforced plastic (FRP); steel, if cathodically protected; a steel-FRP composite; or metal without corrosion protection, if certain soil conditions exist and if approved by the appropriate implementing agency. New piping that routinely contains product must also meet the requirements. Acceptable piping materials include FRP, cathodically protected steel, or unprotected metal, if approved.

This section also addresses spills and overfills of the tank. To prevent releases from such events, spill and overfill equipment must be used. Such equipment could include a catch basin or equivalent device to contain a spill. Overfill prevention devices include automatic shut-off valves, alarms that alert the transfer operator when the tank is close to being full, or flow restrictors on transfer equipment.

In addition, all new tanks and piping must be installed "in accordance with a code of practice developed by a nationally recognized association or independent testing laboratory and in accordance with the manufacturer's instructions." This statement refers to procedures published by well-known organizations like the American Petroleum Institute (API), the Petroleum Equipment Institute (PEI), and the American National Standards Institute (ANSI). While standards of other organizations may also apply, these three organizations have published the most widely followed installation procedures, and tanks and piping installed according to their practices and procedures are usually acceptable.

By December 22, 1998, all existing tank systems are required to be replaced with new systems, permanently closed, or upgraded to meet these new requirements. For existing steel tanks, upgrading may be accomplished by lining the tank, by adding cathodic protection, or by combining these two operations. As with the new tank standards, upgrading must be done in accordance with standards published by a nationally recognized organization such as API, the National Association of Corrosion Engineers (NACE) (cathodic protection), or the National Leak Prevention Association (NLPA) (tank lining). Steel piping must also be cathodically protected. For spill and overfill control, existing tanks must meet the same standards as new tank systems, including catch basins and shut-off valves, flow restrictors, or high-level alarms.

Subpart C: General Operating Requirements [280.30]

Subpart C of the regulations describes requirements and procedures designed to reduce the incidence of releases once tanks are in place and in service. These requirements include spill and overfill controls, corrosion protection, operation and maintenance procedures, ensuring the compatibility of the tank's contents and the tank system's materials, repairs to the tank system, and reporting and recordkeeping for the UST systems.

Many of this subpart's requirements are based on common sense. Such is the case with the spill and overfill requirements, where tank owners and operators are warned to "ensure that the volume available in the tank is greater than the volume to be transferred." Any spills or overfills that occur must be reported, investigated, and cleaned up in accordance with Subpart E [280.53]: Release Reporting, Investigation, and Confirmation.

Systems employing cathodic protection must be inspected regularly by a qualified cathodic protection inspector to ensure the system is operating properly. The cathodic protection system must be inspected within six months, and at least once every three years thereafter. Criteria used will generally be NACE Standard RP-02-85. Records documenting the results of these inspections must be kept.

UST systems found to be failing may be repaired in accordance with the specifications in Subpart C. If the system is constructed of FRP, the repair must be in accordance with the manufacturer's authorized representative or with a nationally recognized repair standard. Metal pipes and fittings cannot be repaired but must be replaced. Repaired steel tanks must be retested within 30 days of the repair to ensure the adequacy of the corrosion protection.

Subpart C also instructs owners and operators to comply fully with inspections, monitoring, and testing performed or required by the implementing agency, as well as with all recordkeeping requirements of the implementing agency. These records must be readily available for inspection (preferably on the site of the UST).

Subpart D: Release Detection [280.40]

Adherence to the performance and operating standards for USTs discussed above will reduce the incidence of leaks from UST systems, both new and existing. The regulations also include extensive requirements for release (leak) detection. EPA views release detection as an essential backup measure combined with prevention techniques. Release detection monitoring on a frequent and consistent basis is the best known method for detecting a release from an UST quickly and reducing the potential environmental damages and liability. Thus, these requirements are in keeping with the overall goal of the UST regulations.

Seven general categories of release detection methods are acceptable:

- tank tightness or precision tests

- manual tank gauging systems

- automatic tank gauging systems

- inventory control methods

- groundwater monitoring

- vapor monitoring

- interstitial monitoring

EPA believes that any of these seven methods can be successful if proper procedures are followed, and thus does not favor one method. Each implementing agency is free to favor one or more of these methods, however, depending on local conditions.

Release detection is required for all UST systems. The deadline for providing detection for existing tank systems varies with the year the tank was installed; older tanks require release detection sooner. New tanks must include release detection upon installation. The detection system must be able to detect leaks from any part of the tank or piping that routinely contains product. The system must be installed, operated, and maintained in accordance with the manufacturer's instructions. The regulations also stipulate performance standards for release detection. Four of the seven methods—inventory control, manual and automatic tank gauging, and tank tightness tests—have specific volume or leak rate limits above which the method must be able to detect a leak with a probability of detection of at least 0.95 (and a probability of false alarm less than 0.05). The other three methods— vapor monitoring, groundwater monitoring, and interstitial monitoring—have no numerical standards to be met, only general standards.

The release detection regulations are divided into petroleum UST systems and hazardous substance UST systems. (All new hazardous substance tank systems must include double-walled tanks.) Generally, UST systems that store petroleum must conduct release detection every 30 days; however, several exceptions apply. New or upgraded USTs may use monthly inventory controls in conjunction with a tank tightness test every five years (until December 22, 1998) or until ten years after installation (for new tanks). Existing USTs (that are not upgraded) may use monthly inventory controls with annual tank tightness tests until December 22, 1998, by which time the tank must be upgraded or closed. Tanks of less than 550 gallons capacity may use weekly tank gauging.

The piping of petroleum UST systems must meet different standards. Pressurized piping must be equipped with an automatic line leak detector and have an annual line tightness test or monthly monitoring. Suction piping does not require release detection, provided certain conditions are met: (1) the below-grade piping operates at negative pressure; (2) the piping is sloped so that product will drain back into the tank; and (3) only one check valve per line is included, and that valve is located as close as practical to the suction pump. Non-exempt suction piping (piping that does not satisfy these conditions) must have a line tightness test at least every three years or monthly monitoring (either vapor, groundwater, or interstitial).

UST systems that store hazardous substances are covered by separate regulations. All existing systems must meet the petroleum release detection requirements described above, and must be upgraded by December 22, 1998, to meet the requirements for new hazardous substance systems. Release detection for new systems (1) must include secondary containment systems that are able to contain all substances released from the tank system until the substances are detected and removed; (2) must prevent the release of any regulated substance to the environment throughout the operational life of the UST system; and (3) must be checked for leakage at least every 30 days. If a double-walled tank is used, it must be able to contain a release from the inner tank within the wall of the outer tank. In such a case, the failure of the inner tank should also be detected. If the system is surrounded by an external liner (e.g., a vault), the external liner must be able to contain the entire leak within its boundary, prevent the interference of groundwater or precipitation with the ability to contain or detect the release, and surround the tank completely (including above the tank). These requirements also pertain to piping for hazardous substance systems. In addition, pressurized piping must be equipped with an automatic line leak detector, similar to petroleum systems.

Release detection systems other than those listed in the regulations may be used provided that the tank owner/operator demonstrates that the alternative method is as effective as a specified method, provides information on corrective action technologies for the stored substances and health risks associated with those substances, and obtains permission from the appropriate implementing agency.

For each of the seven specific types of release detection that may be used to meet the requirements of these regulations (product inventory control, manual tank gauging, tank tightness testing, automatic tank gauging, vapor monitoring, groundwater monitoring, and interstitial monitoring), a performance standard is also specified. For example, if *product inventory control* is used, that method must be able to detect a release of one percent of monthly flow-through, plus 130 gallons, on a monthly basis. If inventory control cannot provide this level of accuracy, it would not qualify as an acceptable release detection method.

The use of *manual tank gauging* is restricted to tanks of 2000 gallons or less. (For tanks with less than 550 gallons capacity, this method alone may satisfy these requirements; for tanks of 550- to 2000-gallon capacity, it can replace inventory control as part of the release detection protocol.) This method's performance standards are specified in terms of a weekly standard and a monthly standard (the average of the four weekly tests). For tanks of 550 gallons or less, the required precision is 10 gallons (weekly) and 5 gallons (monthly). For tanks of 550-1000 gallons, the standards are 13 gallons and 7 gallons. Tanks of 1000-2000 gallons capacity are allowed standards of 26 gallons and 13 gallons, respectively.

A *tank tightness test* must be capable of detecting a leak rate of 0.1 gallons per hour. These tests must also take into account the effects of thermal expansion or contraction of the product, vapor pockets, tank deformation, evaporation or condensation, and the location of the water table.

Automatic tank gauging systems may be used if they can detect a leak rate of 0.2 gallons per hour, and if they are used in conjunction with inventory control.

Vapor monitoring systems are a fifth available release detection method. Prior to their installation, the site must be assessed to ensure that their use is appropriate (i.e., the backfill is sufficiently porous and the stored substance or a tracer is sufficiently volatile to allow detection by the monitoring system, and the background level of contamination will not interfere with the system's operation). These systems must be able to detect any significant increase in vapor level above the background concentration.

Groundwater monitoring may be used as a release detection method; again, this method requires a site assessment. Such an assessment should show that the stored substance is immiscible in water and will float, that the water table is always 20 feet or less below the ground surface, that the soil's hydraulic conductivity is at least 0.01 cm/sec, and that the monitoring wells are designed and placed properly.

Interstitial monitoring may be used for UST systems with secondary containment. Several standard requirements apply, such as the assurance that any leak from the inner tank of a double-walled tank be detected. Specific, detailed requirements are listed in the regulations.

Any other method not listed in the regulations may be used if it can detect a leak rate of 0.2 gallons per hour with a probability of 0.95 (and probability of false alarm less than 0.05), and if such method is approved by the implementing agency.

Subpart E: Release Reporting, Investigation, and Confirmation [280.50]

Direct observation of a leaking UST is impossible because the tank is buried. Therefore, indications that the UST is leaking must be confirmed to establish whether the tank is in fact losing product to the environment. The EPA regulations contain a protocol the tank owner/operators must follow if a tank is suspected of a release. Such suspicions would likely result from positive monitoring results, unusual operating conditions, or a discovery of regulated substances near the tank site. Unusual operating conditions could include a sudden loss of product, the discovery of water in the UST, or erratic behavior of product-dispensing equipment.

The tank owner/operator must report to the implementing agency within 24 hours if a release is suspected. The owner/operator then has three choices: begin corrective action immediately, conduct a tank tightness test, or conduct a site check. Unless the evidence of a release is compelling, corrective action would normally be deferred until the release is confirmed.

The typical method for confirming a suspected release is a tightness test of the tank and piping. If the system test indicates a leak, the UST system must be repaired, removed, or upgraded. Corrective action is also required in such instances. If the tightness test does not indicate that the system is leaking, no further action is necessary, assuming that a release was not suspected due to environmental contamination. If, however, the release was suspected due to environmental contamination (e.g., vapors were detected in a sewer line or free product was found in nearby wells), a site check must be conducted. During the site check, measurements for the presence of regulated substances must be taken where the contaminants are most likely to be found. The owner/operator should consider the nature of the stored substance, the type of initial claim or the cause for suspicion, the type of backfill, depth to groundwater, and other factors when selecting the sampling methods and locations. If the site check fails to indicate a release, no further investigation is required. If a release is confirmed, corrective action must begin.

Subpart F: Release Response and Corrective Action [280.60]

This section describes activities to investigate, report, abate, and remedy releases from USTs into the environment. Such actions would follow the confirmation of a release (either in accordance with the procedures described in Subpart E or otherwise). The prescribed response to a confirmed release includes two phases: immediate actions to identify and reduce imminent health threats, such as fire or explosion; and actions to mitigate long-term threats to human health and the environment. Immediate actions might include pumping the remaining product from a leaking tank or dispersing explosive vapors, while longer-term corrective actions could include groundwater cleanup plans using air stripping or other such measures.

Subpart F contains seven subsections detailing actions to be taken following a release confirmation:

- the initial response
- initial abatement measures and site check
- an initial site characterization
- free product removal
- investigations for soil and groundwater cleanup
- the corrective action plan
- public participation

The *initial response* to any confirmed release should include three steps: notification of the release to the implementing agency; prevention of any further releases of regulated substances to the environment; and the mitigation of any immediate fire, explosion, or vapor hazards. These steps must all occur within 24 hours of the confirmation of the release.

Following the initial response, the owner/operator should immediately begin *initial abatement measures*. The following steps have been enumerated by EPA to satisfy this requirement: (1) remove regulated substance from the UST to prevent further release; (2) visually inspect any exposed portion of the release and prevent further migration; (3) continue to mitigate potential fire and explosion threats; (4) remedy hazards that result from exposed contaminated soil; and (5) determine whether free product is present, and if so, begin removal procedures. Within 20 days after the release confirmation, owners and operators must submit a report to the implementing agency summarizing the steps taken.

A *site characterization* must also be performed and a report submitted to the implementing agency within 45 days of release confirmation. This report should contain information gained while confirming the release or completing the initial abatement measures, as well as other information about the nature and quantity of the release, surrounding populations, location and use of nearby wells, land use, climatological conditions, and similar factors.

At sites where the initial site check determined the presence of free product, steps must be taken to *remove free product* "to the maximum extent practicable." At the minimum, this procedure should strive to abate further migration of the free product and should ensure that flammable products are handled in a safe and competent manner. A report detailing the free product removal should be submitted within 45 days of release confirmation to the implementing agency. The report must include information regarding the quantity and type of free product encountered, the type of recovery system being used, the name of the person performing the free product recovery operation, and information on the treatment and ultimate disposition of the recovered product.

Investigations for soil and groundwater cleanup must be conducted if onsite conditions warrant. These conditions include the presence of free product, contamination of nearby wells, or evidence that contaminated soil is in contact with groundwater. The information collected during this investigation must be submitted to the implementing agency "as soon as practicable," although there is no specified time limit.

After reviewing the information gained up to this time, the implementing agency may require the owner and operator to submit a *corrective action plan* for soil and groundwater cleanup. The plan will be approved only after the implementing agency determines that it adequately protects human health and the environment, taking into account the characteristics (e.g., toxicity) of the regulated substance, hydrogeological conditions of the site, exposure assessment, and proximity of nearby water sources. In the interest of minimizing environmental contamination, cleanup may begin prior to the approval of the implementing agency.

For all confirmed releases that require a corrective action plan, the implementing agency must take a number of steps to assure *public participation* and notice of the corrective action procedure. These steps could include, for example, public notice in newspapers or letters to affected households. A public meeting may also be held to consider comments on the corrective action plan.

Subpart G: Out-of-Service UST Systems and Closures [280.70]

This section describes requirements for the closure of a UST system. The primary objectives of these requirements are to identify and contain existing contamination and to prevent future contamination from USTs that are no longer in service. Proper management of out-of-service USTs will become increasingly important during the next decade, when large numbers of USTs will likely be taken out of service in response to these regulations. This section also discusses procedures to follow if a tank will be out of service only temporarily. Specific requirements of this subpart are discussed below.

If a UST system is *temporarily closed*, the corrosion protection and release detection systems must be maintained and continue in operation. (Release detection need not be operated if the UST system is empty.) If the closure exceeds three months, access from manways must be closed off, but vent lines must be left open and operating. A UST that is temporarily closed for more than 12 months must be closed permanently if it does not meet standards for new or upgraded USTs.

For tanks undergoing *permanent closure or a change in service*, the implementing agency must be notified and a site assessment must be performed. If the tank is to be closed, the UST should be emptied, cleaned, and removed, or filled with an inert (solid) material. Tanks undergoing a change in service (i.e., to store nonregulated substances) must also first be emptied and cleaned. Prior to either of these actions, a site assessment must be performed. If contaminated soil, contaminated groundwater, or free product is found, corrective action must begin in accordance with Subpart F.

Owners and operators of *USTs closed permanently prior to De-*

cember 22, 1988 must at the direction of the implementing agency perform a site assessment.

Records must be maintained that demonstrate compliance with the closure requirements in this section.

Summary

The preceding sections summarize the most important facets of the UST regulations. It should be noted that this summary is not complete; where further information is desired or where legal issues are involved, the regulation itself (40 CFR 280 or *Federal Register,* September 23, 1988) should be consulted. For a thorough discussion of EPA's decisionmaking process, readers may consult the text of the proposed rule (*Federal Register,* April 17, 1987) or the preamble of the final rule (*Federal Register,* September 23, 1988). Figure 36 is a graphical summary of the rule as discussed in this chapter, and may be useful as a source of quick reference.

DEFINITIONS

The following terms and their definitions are adapted from the UST regulations. This list should be helpful in understanding the preceding summary of the regulations, as well as being a good reference for the future.

Above-ground release   any release to the surface of the land or to surface water. This includes, but is not limited to, release from the above-ground portion of an UST system and above-ground releases associated with overfills and transfer operations as the regulated substances move to or from an UST system.

Ancillary equipment   any devices including, but not limited to, such devices as piping, fittings, flanges, valves, and pumps used to distribute, meter, or control the flow of regulated substances to and from an UST.

Below-ground release   any release to the subsurface of the land and to groundwater. This includes, but is not limited to, releases from the below-ground portions of an underground storage tank system and below-ground releases associated with overfills and transfer opera-

Figure 36.   Federal UST regulations. Source: Camp Dresser & McKee.

Reg. Citation		Compliance Requirements	Reporting/Record Keeping
Subpart B: Performance Standards			
280.20(a)	NEW TANK OPTIONS • FRP or • Steel with CP or • Steel-FRP composite or • Metal without CP if approved	• Design, construction, and CP must be in accordance with code(s). Compliance with CP must be documented.	• UST notification form to IA (within 30 days) • Certificate of Compliance for CP; CP analysis required if CP not used
280.20(b)	NEW PIPING OPTIONS • FRP or • Steel with CP or • Metal without CP (if approved)		
280.20(c)	SPILL PREVENTION • Catch basin or equivalent		
280.20(c)	OVERFILL PREVENTION OPTIONS • Automatic shut-off valves or • Flow restrictors or • High-level alarm		
280.20(d)	INSTALLATION • Must be in accordance with code and manufacturer's instructions	• Compliance with installation must be documented	• Certificate of Compliance for installation
280.21(b)	UPGRADING EXISTING TANKS • Interior lining and/or CP	• By December 22, 1998 all UST systems must be replaced, upgraded, or closed	
280.21(c)	UPGRADING EXISTING PIPING • CP	• Installation must be in accordance with code(s) and inspections as per regulations	
280.21(d)	UPGRADING FOR SPILL AND OVERFILL PREVENTION • Same as for new systems		
Subpart C: Operating Requirements			
280.30	SPILL AND OVERFILL CONTROL • If spill occurs ──► GO TO SUBPART E		• Report to IA
280.31	OPERATION & MAINTENANCE OF CP • Must be inspected by qualified tester	• Within 6 months of installation, then every 3 years • If impressed current CP inspect every 60 days	• Maintain test results of last 2 inspections • Maintain test results of last 3 inspections
280.32	COMPATIBILITY with stored substance required		
280.33	REPAIRS: • FRP tanks and fiberglass pipes/fittings • Metal pipes/fittings that have released product • CP	• Repair as per code(s) or mfgr. and tightness test • Must replace • Repair and test	• Maintain records of compliance
Subpart D: Release Detection			
280.40	ALL USTS MUST HAVE RD	• Within phase-in period; else UST must be closed	
	• Performance requirements must be met	• Methods used after December 22, 1990 must detect at P_d=0.95 and P_{fa}=0.05 for that method	• Submit performance claims and how determined; maintain records for 5 years
	• Installation, calibration, O & M, repair	• As per manufacturer's instructions	• Maintain documentation of servicing for 1 year • Maintain mfgr. schedules of required servicing for 5 years
	• Sampling, testing, monitoring	• As per regulations	• Maintain results for 1 year
	IF RD INDICATES A LEAK ──► GO TO SUBPART E		• Report to IA
280.41	**REQUIREMENTS FOR PETROLEUM UST SYSTEMS** TANKS—MONITOR EVERY 30 DAYS EXCEPT: • New/upgraded: Monthly inventory controls & TTT	• TTT: every 5 years until December 22, 1998 or until 10 years after installation/upgrade	

Figure 36, continued.

Reg. Citation	Compliance Requirements	Reporting/Record Keeping
	• Existing: monthly inventory controls & TTT	• TTT: annually until December 22, 1998 when upgrade is required
	• If UST is less than 550 gallons: tank gauging allowed	• Weekly
280.41(b)	PIPING—PRESSURIZED MUST HAVE: • Automatic line leak detection and line TT or monitoring	• Line TT: annual; monitoring (vapor, groundwater, interstitial): monthly
280.41(b)	PIPING—SUCTION MUST HAVE: • Line TT or monitoring	• Line TT: every 3 years; monitoring (vapor, groundwater, interstitial): monthly
280.42	**REQUIREMENTS FOR HAZARDOUS SUBSTANCE UST SYSTEMS**	
	EXISTING SYSTEMS—MEET REQUIREMENTS FOR PETROLEUM UST SYSTEMS	• By December 22, 1998 must meet RD for new hazardous systems (interstitial monitoring)
	NEW SYSTEMS MUST HAVE RD FOR: • Secondary containment systems • Double-walled tanks • External liners • Piping with secondary containment	• Several requirements apply, see regs
280.43	**RD METHODS ALLOWED**	
	TANKS—MONTHLY METHODS:	PERFORMANCE STANDARDS:
280.43(d)	• ATGS	• 0.2 gal/hr with inventory control requirements
280.43(e)	• Vapor monitoring (requires site assessment)	• Any significant increase in concentration above background
280.43(f)	• Groundwater monitoring (requires site assessment)	• 1/8" of free product on groundwater table
280.43(g)	• Interstitial monitoring	• Standard requirements apply
280.43(h)	• Other methods	• 0.2 gal/hr or 150 gallons within a month
	TANKS—OTHER METHODS:	
280.43(a)	• Inventory controls (measure daily, reconcile monthly)	• 1% of flow-through plus 130 gal monthly
280.43(b)	• Manual tank gauging (weekly or monthly) Only for tanks ≤ 2000 gal nominal capacity	• If tank ≤ 550 gal, detect 10 gal/wk or 5 gal/mo (may use this as sole method of RD) If tank > 550 and ≤ 1000 gal, detect 13 gal/wk or 7 gal/mo (may use in place of manual inventory controls) If tank > 1000 and ≤ 2000 gal, detect 26 gal/wk or 13 gal/mo (may use in place of manual inventory controls)
280.43(c)	• Tank tightness testing (frequency: see above §280.41)	• 0.1 gal/hr and account for various effects
280.44	PIPING—(frequency: see above §280.41)	
280.44(a)	• Automatic line leak detectors	• 3 gal/hr at 10 psi within 1 hour
280.44(b)	• Line tightness testing	• 0.1 gal/hr at 1.5 times operating pressure
280.44(c)	• Vapor, groundwater, or interstitial monitoring	• See above

Subpart E: Release Reporting, Investigation, Confirmation

Reg. Citation	Compliance Requirements	Reporting/Record Keeping
280.50	SUSPECTED RELEASES IF: • Discover leak at/near site or • Observe unusual operating conditions (and equipment is not defective) or • RD indicates leak (and equipment is not defective and additional monitoring is negative) or • Off-site impacts might indicate a leak	• Report within 24 hours
280.51	THEN: Investigate & confirm: • Conduct tightness tests on system:	
280.52		• Within 7 days unless CA is initiated

IF: • Leak → repair, replace or upgrade → GO TO SUBPART F
• No leak and no contamination → STOP
• No leak and contamination → DO SITE CHECK
• Conduct site check (excavation zone & site):
IF: • Leak → GO TO SUBPART F
• No leak → STOP

280.53	SPILLS AND OVERFILLS—CONTAIN AND CLEANUP and	• If cleanup of releases of < 25 gal (petroleum) or < RQ (hazardous substance) cannot be accomplished in 24 hours, must notify agency	
	IF: • Release> 25 gal petroleum or creates sheen		• Report within 24 hours
	• Or if release> RQ (under CERCLA) → GO TO SUBPART F		• Report immediately

Subpart F: Release Response and Corrective Action

280.61	INITIAL RESPONSE: Stop leak, mitigate hazards	• Within 24 hours — mandatory	
280.62	INITIAL ABATEMENT MEASURES AND SITE CHECK	• Mandatory	• Report confirmation within 24 hours
280.63	INITIAL SITE CHARACTERIZATION: Assemble information	• Mandatory	• Report progress within 20 days
280.64	FREE PRODUCT REMOVAL: Abate migration		• Submit information within 45 days
280.65	INVESTIGATIONS FOR SOIL & GROUNDWATER CLEANUP	• If on-site conditions warrant	• Submit report within 45 days
			• Submit information when practicable
280.66	CORRECTIVE ACTION PLAN:		
	• IA sets schedule and goals	• At direction of IA	• Submit per IA schedule
	• Owner/operator establishes strategy	• Upon approval, implement plan	
	• IA notifies affected public		

Subpart G: Out-of-Service UST Systems and Closure

280.70	TEMPORARY REMOVAL FROM USE: MUST CONTINUE O&M OF CP AND RD UNLESS UST IS EMPTY		
	• If out-of-service more than 3 months → close off access		
	• If out-of-service more than 12 months → permanent closure unless it has maintained CP and RD		
280.71	PERMANENT CLOSURE AND CHANGES IN SERVICE: MUST PERFORM SITE ASSESSEMENT		
	• If closing: empty, clean, and remove or fill with inert solid material		• Maintain records of compliance
	• If change in service: empty, clean, and store non-regulated substance		• Notify 30 days prior to closure
			• Notify 30 days prior to change in service
280.72	SITE ASSESSMENT MEASURE FOR RELEASES	• External RD meets requirement	• Maintain assessment results for 3 years
	• If release discovered → GO TO SUBPART F		
280.73	UST SYSTEMS CLOSED PRIOR TO DECEMBER 22, 1988: Perform site assessment and close per regs	• At agency direction	

Abbreviations

ATGS Automatic tank gauging system
CA Corrective action
CP Corrosion protection
Code Code of practice developed by a nationally recognized association or independent testing laboratory (specified throughout regulations)

FRP Fiberglass-reinforced plastic
IA Implementing agency
MFGR Manufacturer
O&M Operation and maintenance
P_D Probability of detection
P_{FA} Probability of false alarm
RD Release detection
RQ Reportable quantity (of hazardous substance under CERCLA)

TTT Tank tightness test
TT Tightness test
UST Underground storage tank

Please note: This chart is not exhaustive. Many exceptions apply. Specific requirements are detailed in the regulations—see regulatory citation numbers for applicable section of the final regulations as published in the Federal Register on September 23, 1988 (40 CFR Part 280).

tions as the regulated substances move to or from an underground storage tank.

Beneath the surface of the ground   beneath the ground surface or otherwise covered with earthen materials.

Cathodic protection   a technique to prevent corrosion of a metal surface by making that surface the cathode of an electrochemical cell. For example, a tank system can be cathodically protected through the application of either galvanic anodes or impressed current.

Cathodic protection tester   a person who can demonstrate an understanding of the principles and measurements of all common types of cathodic protection systems as applied to buried or submerged metal piping and tank systems. At a minimum, such persons must have education and experience in soil resistivity, stray current, structure-to-soil potential, and component electrical isolation measurements of buried metal piping and tank systems.

CERCLA   the Comprehensive Environmental Response, Compensation, and Liability Act of 1980, as amended.

Compatible   the ability of two or more substances to maintain their respective physical and chemical properties upon contact with one another for the design life of the tank system under conditions likely to be encountered in a UST.

Connected piping   all underground piping including valves, elbows, joints, flanges, and flexible connectors attached to a tank system through which regulated substances flow. For the purpose of determining how much piping is connected to any individual UST system, the piping that joins two UST systems should be allocated equally between them.

Consumptive use   consumed on the premises. This term is used with respect to heating oil.

Corrosion expert   a person who, by reason of thorough knowledge of the physical sciences and the principles of engineering and mathematics acquired by a professional education and related practical experience, is qualified to engage in the practice of corrosion control on buried or submerged metal piping systems and metal tanks. Such a person must be accredited or certified as being qualified by the National Association of Corrosion Engineers or be a registered professional engineer who has certification or licensing that includes education and experience in corrosion control of buried or submerged metal piping systems and metal tanks.

Dielectric material a material that does not conduct direct electrical current. Dielectric coatings are used to electrically isolate UST systems from the surrounding soils. Dielectric bushings are used to electrically isolate portions of the UST system from each other (e.g., tank from piping).

Electrical equipment underground equipment containing dielectric fluid that is necessary for the operation of equipment such as transformers and buried electrical cable.

Excavation zone the volume containing the tank system and backfill material bounded by the ground surface, walls, and floor of the pit and trenches into which the UST system is placed at the time of installation.

Existing tank system a tank system used to contain an accumulation of regulated substances or for which installation has commenced on or before December 22, 1988. Installation is considered to have commenced if the owner or operator has obtained all federal, state, and local approvals or permits necessary to begin physical construction of the site or installation of the tank system; *and either* (1) if a continuous onsite physical construction or installation program has begun; *or* (2) if the owner or operators have entered into contractual obligations—which cannot be canceled or modified without substantial loss—for physical construction at the site or installation of the tank system to be completed within a reasonable time.

Farm tank a tank located on a tract of land devoted to the production of crops and raising animals, including fish, and associated residences and improvements. A farm tank must be located on the farm property. "Farm" includes fish hatcheries, rangelands, and nurseries with growing operations.

Flow-through process tank a tank that forms an integral part of a production process through which there is a steady, variable, recurring, or intermittent flow of materials during the operation of the process. Flow-through process tanks do not include tanks used for storage of materials prior to their introduction into the production process or for the storage of finished products or by-products from the production process.

Free product a regulated substance that is present as a non-aqueous phase liquid (i.e., liquid not dissolved in water.)

Gathering lines   any pipeline, equipment, facility, or building used in the transportation of oil or gas during oil or gas production or gathering operations.

Hazardous substance UST system   an underground storage tank system that (1) contains a hazardous substance defined in section 101(14) of the Comprehensive Environmental Response, Compensation, and Liability Act of 1980 (but not including any substance regulated as a hazardous waste under subtitle C) or any mixture of such substances and petroleum, *and* (2) is not a petroleum UST system.

Heating oil   petroleum whose fuel oil technical grade is No. 1, No. 2, No. 4—light, No. 4—heavy, No. 5—light, No. 5—heavy, or No. 6; other residual fuel oils (including Navy Special Fuel Oil and Bunker C); and other fuels when used as substitutes for one of these fuel oils. Heating oil is typically used in the operation of heating equipment, boilers, or furnaces.

Hydraulic lift tank   a tank holding hydraulic fluid for a closed-loop mechanical system that uses compressed air or hydraulic fluid to operate lifts, elevators, and other similar devices.

Implementing agency   EPA or, in the case of a state with a program approved under section 9004 (or pursuant to a memorandum of agreement with EPA), the designated state or local agency responsible for carrying out an approved UST program.

Liquid trap   sumps, well cellars, and other traps used in association with oil and gas production, gathering, and extraction operations (including gas production plants) for the purpose of collecting oil, water, and other liquids. These liquid traps may temporarily collect liquids for subsequent disposition or reinjection into a production or pipeline stream, or may collect and separate liquids from a gas stream.

Maintenance   the normal operational upkeep to prevent an underground storage tank system from releasing product.

Motor fuel   petroleum or a petroleum-based substance that is motor gasoline, aviation gasoline, No. 1 or No. 2 diesel fuel, or any grade of gasohol, and is typically used in the operation of a motor engine.

New tank system   a tank system that will be used to contain an accumulation of regulated substances and for which installation has commenced after December 22, 1988. (See also "Existing tank systems.")

Noncommercial purposes not for resale. This term is used with respect to motor fuel.

On the premises where stored with respect to heating oil, UST systems located on the same property where the stored heating oil is used.

Operational life the period beginning when installation of the tank system has commenced until the time the tank system is properly closed under Subpart G.

Operator any person in control of, or having responsibility for, the daily operation of the UST system.

Overfill release a release that occurs when a tank is filled beyond its capacity, resulting in a discharge of the regulated substance to the environment.

Owner (a) in the case of an UST system in use on November 8, 1984, or brought into use after that date, any person who owns an UST system used for storage, use, or dispensing of regulated substances; (b) in the case of any UST system in use before November 8, 1984, but no longer in use on that date, any person who owned such UST immediately before the discontinuation of its use.

Person an individual, trust, firm, joint stock company, federal agency, corporation, state, municipality, commission, political subdivision of a state, or any interstate body. "Person" also includes a consortium, a joint venture, a commercial entity, and the United States Government.

Petroleum UST system an underground storage tank system that contains petroleum or a mixture of petroleum with *de minimis* quantities of other regulated substances. Such systems include those containing motor fuels, jet fuels, distillate fuel oils, residual fuel oils, lubricants, petroleum solvents, and used oils.

Pipe or **Piping** a hollow cylinder or tubular conduit that is constructed of non-earthen materials.

Pipeline facilities (including **gathering lines**) new and existing pipe rights-of-way and any associated equipment, facilities, or buildings.

Regulated substance (a) any substance defined in section 101(14) of the Comprehensive Environmental Response, Compensation, and Liability Act of 1980 (CERCLA) (but not including any substance regulated as a hazardous waste under subtitle C); and (b) petroleum, including crude oil or any fraction thereof that is liquid at standard conditions of temperature and pressure (60°F and 14.7 psia). The term "regulated substance" includes but is not limited to petroleum and petroleum-based substances, comprised of a complex blend of hydrocarbons derived from crude oil though processes of separation, conversion, upgrading, and finishing, such as motor fuels, jet fuels, distillate fuel oils, residual fuel oils, lubricants, petroleum solvents, and used oils.

Release   any spilling, leaking, emitting, discharging, escaping, leaching, or disposing from an UST into groundwater, surface water or subsurface soils.

Release detection   determining whether a release of a regulated substance has occurred from the UST system into the environment or into the interstitial space between the UST system and its secondary barrier or surrounding secondary containment.

Repair   to restore a tank or UST system component that has caused a release of product from the UST system.

Residential tank   a tank located on property used primarily for dwelling purposes.

SARA   the Superfund Amendments and Reauthorization Act of 1986.

Septic tank   a water-tight covered receptacle designed to receive or process, through liquid separation or biological digestion, the sewage discharged from a building sewer. The effluent from such a receptacle is distributed for disposal though the soil and settled solids and scum from the tank are pumped out periodically and hauled to a treatment facility.

Storm water or wastewater collection system   piping, pumps, conduits, and any other equipment necessary to collect and transport the flow of (1) surface water runoff resulting from precipitation, or (2) domestic, commercial, or industrial wastewater to and from retention areas or any areas where treatment is designated to occur. The collection of storm water and wastewater does not include treatment except where incidental to conveyance.

Surface impoundment   a natural topographic depression, manmade excavation, or diked area formed primarily of earthen materials (although it may be lined with manmade materials) that is not an injection well.

Tank   a stationary device designed to contain an accumulation of regulated substances and constructed of non-earthen materials (e.g., concrete, steel, plastic) that provide structural support.

Tank system   see "UST system."

Underground area   an underground room, such as a basement, cellar, shaft, or vault, providing enough space for physical inspection of the exterior of the tank situated on or above the surface of the floor.

Underground release   any below-ground release.

Underground storage tank or **UST**   any one or combination of tanks (including underground pipes connected thereto) used to contain an accumulation of regulated substances, the volume of which (including the volume of underground pipes connected thereto) is 10% or more beneath the surface of the ground. This term does *not* include the following types of storage vessels (or any pipes connected thereto):

- any farm or residential tank of 1100 gallons or less capacity used for storing motor fuel for noncommercial purposes

- any tank used for storing heating oil for consumptive use on the premises where stored

- any septic tank

- any pipeline facility (including gathering lines) that is (1) regulated under the Natural Gas Pipeline Safety Act of 1968 (49 U.S.C. App. 1671 et. seq.) or the Hazardous Liquid Pipeline Safety Act of 1979 (49 U.S.C. App. 2001 et seq.); *or* (2) an intrastate pipeline facility regulated under state laws comparable to the provisions of the above-mentioned federal laws

- any surface impoundment, pit, pond, or lagoon

- any storm water or wastewater collection system

- any flow-through process tank

- any liquid trap or associated gathering lines directly related to oil or gas production and gathering operations

- any storage tank situated in an underground area (such as a basement, cellar, mineworking, drift, shaft, or tunnel) if the storage tank is situated on or above the surface of the floor

Upgrade   the addition or retrofit of systems such as cathodic protection, lining, or spill and overfill controls to improve the ability of an underground storage tank system to prevent the release of product.

UST system   an underground storage tank, connected underground piping, underground ancillary equipment, and containment system, if any.

Wastewater treatment tank   a tank that is designed to receive and treat an influent wastewater through physical, chemical, or biological methods.

APPENDIX B   BIBLIOGRAPHY

Alexander, M. 1985. "Biodegradation of Organic Chemicals." *Env. Sci. Tech.* 8(2):106-111.

Amdurer, M., R. T. Fellman, J. Roetzer, and C. Russ. 1986. "Systems to Accelerate In-situ Stabilization of Waste Deposits." U.S. EPA, EPA/590/2-86/002.

Ball, W. P., M. D. Jones, and M. C. Kavanaugh. 1984. "Mass Transfer of Volatile Organic Compounds in Packed Tower Aeration." *J. Water Pollut. Control Fed.* 56(2):127-136.

Barker, J. F., G. C. Patrick, and D. Major. 1987. "Natural Attenuation of Aromatic Hydrocarbons in a Shallow Sand Aquifer." *Ground Water Mon. Rev.* 7(1):64-71.

Bossert, I., and R. Bartha. 1984. "The Fate of Petroleum in Soil Ecosystems." In *Petroleum Microbiology*, R. M. Atlas, Ed. (New York: MacMillan Publishing Co., Inc.).

Bourdeau, R. 1987. Calgon Carbon Corporation. Personal communication with J. Curtis (Camp Dresser & McKee Inc., Boston).

Bouwer, E. J., and P. W. McCarty. 1984. "Modeling of Trace Organics Biotransformation in the Subsurface." *Ground Water* July-August, pp. 433-440.

Bright, R. L., and M. H. Stenzel. 1985. "Contaminated Groundwater Treatment Using Granular Activated Carbon and Air Stripping." For presentation to Committee on Refinery Environmental Control, American Petroleum Institute.

Brunotts, V. A., L. R. Emerson, E. N. Rebis, and A. J. Roy. 1983. "Cost Effective Treatment of Priority Pollutant Compounds with Granular Activated Carbon." In *National Conference on Management of Uncontrolled Hazardous Waste Sites* (Silver Spring, MD: Hazardous Materials Control Research Institute), pp. 209-216.

Camp Dresser & McKee Inc. 1986. "Special Water Treatment Study— Phase II" (Draft). Prepared for New Jersey Department of Environmental Protection.

Camp Dresser & McKee Inc. 1987a. "Special Water Treatment Feasibility Study—Phase II." Workshop on Response to Volatile Organic Chemicals in Public Water Supplies. Prepared for New Jersey Department of Environmental Protection, Division of Water Resources.

Camp Dresser & McKee Inc. 1987b. "New Jersey Special Water Treatability Study—Phase II." Prepared for New Jersey Department of Environmental Protection.

Cerniglia, C. E. 1984. "Microbial Transformation of Aromatic Hydrocarbons." In *Petroleum Microbiology,* R. M. Atlas, Ed. (New York: MacMillan Publishing Co., Inc.), pp. 99-128.

Chan, D. B., and E. A. Ford. 1986. "In-situ Oil Biodegradation." *The Military Engineer* 509:447-449.

Cheremisinoff, P. N., and F. Ellerbusch. 1978. *Carbon Adsorption Handbook* (Ann Arbor, MI: Ann Arbor Science Publishers, Inc.).

Chrobak, R. S., D. L. Kelleher, and I. H. Suffet. 1985. "Full Scale GAC Adsorption Performance Compared to Pilot Plant Predictions." In *Proceedings ASCE Environmental Engineering Conference* (New York: American Society of Civil Engineers), pp. 1115-1150.

Clark, R. M., R. G. Eilers, and J. A. Goodrich. 1984. "VOCs in Drinking Water: Cost of Removal." *J. Environ. Eng.* 110(6):1146-1162.

Cookson, J. T., Jr. 1978. "Adsorption Mechanisms: The Chemistry of Organic Adsorption on Activated Carbon." In *Carbon Adsorption Handbook*, P. N. Cheremisinoff and F. Ellerbusch, Eds. (Ann Arbor, MI: Ann Arbor Science Publishers), pp. 241-279.

Cooney, J. J. 1984. "The Fate of Petroleum Pollutants in Freshwater Ecosystems." In *Petroleum Microbiology,* R. M. Atlas, Ed. (New York: MacMillan Publishing Co., Inc.) pp. 399-433.

Crittenden, J. C., D. W. Hand, H. Arora, and B. W. Lykins, Jr. 1987. "Design Considerations for GAC Treatment of Organic Chemicals." *J. Am. Water Works Assoc.* 79(1):74-82.

Cummins, M. D., and J. J. Westrick. 1982. "Packed Column Air Stripping for Removal of Volatile Compounds." In *Proceedings ASCE Environmental Engineering Conference* (New York: American Society of Civil Engineers), pp. 571-585.

Cummins, M. D., and J. J. Westrick. 1983. "Trichloroethylene Removal by Packed Column Air Stripping: Field Verified Design Procedure." In *Proceedings ASCE Environmental Engineering Conference* (Boulder, CO: American Society of Civil Engineers), pp. 442-449.

Dobbs, R. A., and J. M. Cohen. 1980. "Carbon Adsorption Isotherms for Toxic Organics." U.S. EPA, Office of Research and Development. Cincinnati, Ohio.

Dyksen, J. E., A. F. Hess, M. J. Barnes, and G. C. Cline. 1982. "The Use of Aeration to Remove Volatile Organics from Groundwater." Paper presented at the 1982 Annual Conference of the American Water Works Association, Miami Beach, FL, May 1982.

Eckert, J. S., et al. 1970. "Selecting the Proper Distillation Column Packing." *Chem. Eng. Prog.* 66(3):39-44.

Ehrenfeld, J., and J. Bass. 1984. "Evaluation of Remedial Action Unit Operations at Hazardous Waste Disposal Sites." *Poll. Tech. Rev.* 110:65-114.

Engineering-Science Inc. 1986. "Cost Model for Selected Technologies for Removal of Gasoline Components from Groundwater." Prepared for the American Petroleum Institute, Washington, DC.

Garrett, P., M. Moreau, and J. D. Lowry. 1986. "Methyl Tertiary Butyl Ether as a Ground Water Contaminant." Proceedings of the Petroleum Hydrocarbons and Organic Chemicals in Ground Water Conference. National Water Well Association–American Petroleum Institute.

Gumerman, R. C., R. L. Culp, and R. M. Clark. 1979. "The Cost of Granular Activated Carbon Adsorption Treatment." *J. Am. Water Works Assoc.* 71(11):690-696.

Guisti, D. M., R. A. Conway, and C. T. Lawson. 1974. "Activated Carbon Adsorption of Petrochemicals." *J. Water Pollut. Control Fed.* 46(5):947-965.

Hall, D. W., and R. L. Mumford. 1987. "Interim Private Water Well Remediation Using Carbon Adsorption." *Ground Water Mon. Rev.* 7(1):77-83.

Hand, D. W., J. C. Crittenden, J. L. Gehin, and B. W. Lykins, Jr. 1986. "Design and Evaluation of an Air Stripping Tower for Removing VOCs from Groundwater." *J. Am. Water Works Assoc.* 78(9):87-97.

Hess, A. F., J. E. Dyksen, and G. C. Cline. 1981. "Case Studies Involving Removal of Organic Chemical Compounds from Groundwater." In *Organic Chemical Contaminants in Groundwater: Transport and Removal,* pp. 965-980. American Water Works Association Seminar Proceedings, June 7, 1981.

Hess, A. F., J. E. Dyksen, and H. J. Dunn. 1983. "Control Strategy— Aeration Treatment Techniques." In *Occurrence and Removal of Volatile Organic Chemicals from Drinking Water* (Denver, CO: American Water Works Association Research Foundation), pp. 87-155.

Horvath, R. S. 1972. "Microbial Co-Metabolism and the Degradation of Organic Compounds in Nature." *Bacteriol. Rev.* 36(2):146-166.

ICF Incorporated. 1985. Superfund Public Health Evaluation Manual (Draft). Prepared for U.S. EPA, Office of Solid Waste and Emergency Response.

Jolley, R. L. 1981. "Concentrating Organics in Water for Biological Testing." *Env. Sci. Technol.* 15(8):874-880.

Kavanaugh, M. C., and R. R. Trussell. 1980. "Design of Aeration Towers to Strip Volatile Contaminants from Drinking Water." *J. Am. Water Works Assoc.* 72(12):684-691.

Kavanaugh, M. C., and R. R. Trussell. 1981. "Air Stripping as a Treatment Process." In *Organic Chemical Contaminants in Groundwater: Transport and Removal,* pp. 83-106. American Water Works Association Seminar Proceedings, June 7, 1981.

Knox, R. C., L. W. Canter, D. J. Kincannon, and E. L. Stover. 1984. *State-of-the-Art Aquifer Restoration,* Vol. II, Appendices A through G. National Center for Ground Water Research. Prepared for U.S. EPA.

Love, O. T., Jr., and R. G. Eilers. 1982. "Treatment of Drinking Water Containing Trichloroethylene and Related Industrial Solvents." *J. Am. Water Works Assoc.* 76(8):413-425.

MacLeod, J. E., and G. R. Allan. 1983. "Activated Carbon Treatment Restores Acton Water Supply." *American City & County* 98(11):32-35.

Matthews, J. 1987. U.S. EPA (Robert S. Kerr Environmental Research Laboratory, Ada, OK). Personal communication with T. Pedersen (Camp Dresser & McKee Inc., Boston), February 20.

Mattson, J. S., and H. B. Mark, Jr. 1971. *Activated Carbon: Surface Chemistry and Adsorption from Solution* (New York: Marcel Dekker, Inc.)

McCarty, P. L., D. Argo, and M. Reinhard. 1979. "Operational Experiences with Activated Carbon Adsorbers at Water Factory 21." *J. Am. Water Works Assoc.* 71(11):683-689.

McGuire, M. J., and I. H. Suffet. 1983. *Treatment of Water by Activated Carbon* (Washington, DC: American Chemical Society).

McIntyre, G. T., N. H. Hatch, S. R. Gelman, and T. J. Peschman. 1986. "Design and Performance of a Groundwater Treatment System for Toxic Organics Removal." *J. Water Pollut. Control Fed.* 58(1):41-46.

McKinnon, R. J., and J. E. Dyksen. 1984. "Removing Organics from Groundwater Through Aeration Plus GAC." *J. Am. Water Works Assoc.* 76(5):42-47.

Medlar, S. Camp Dresser & McKee Inc. 1987. Presentation at Workshop on Response to Volatile Organic Chemicals in Public Water Supplies. Edison, NJ, March 19.

Munz, C., and P. V. Roberts. 1987. "Air-Water Phase Equilibria of Volatile Organic Solutes." *J. Am. Water Works Assoc.* 79(5):62-69.

Nirmalakhandan, N., Y. H. Lee, and R. E. Speece. 1987. "Designing a Cost-Effective Air Stripping Process." *J. Am. Water Works Assoc.* 79(1):56-63.

Neulight, J. Calgon Carbon Corporation. 1987. Personal communication with J. Curtis (Camp Dresser & McKee Inc., Boston), March 23.

Nyer, E. K. 1985. *Groundwater Treatment Technology* (New York: Van Nostrand Reinhold Company).

O'Brien, R. P., and J. L. Fisher. 1983. "There Is an Answer to Groundwater Contamination." *Water Engineering & Management* 130(5): 30-70.

O'Brien, R. P., and M. H. Stenzel. 1984. "Combining Granular Activated Carbon and Air Stripping." *Public Works* 115(12):54-62.

Onda, K., H. Takeuchi, and Y. Okumoto. 1968. "Mass Transfer Coefficients Between Gas and Liquid Phases in Packed Columns." *J.
Chem. Eng. Japan* 72(12):684.

Perry, J. J. 1984. "Microbial Metabolism of Cyclic Alkanes." In *Petroleum Microbiology*, R. M. Atlas, Ed. (New York: MacMillan Publishing Co., Inc.), pp. 61-68.

Perry, R. H., and C. H. Chilton. 1973. *Chemical Engineer's Handbook*,
5th ed. (New York: McGraw-Hill Book Company).

Polybac. 1980. Technical Data Sheet 805B, Allentown, PA.

Polybac. 1981. Technical Data Sheet 818N, Allentown, PA.

Raczko, R. F., J. E. Dyksen, and M. B. Denove. 1984. "Air Stripping for
Removal of Volatile Organics from Groundwater: From Pilot
Studies to Full-Scale Systems." *Clearwaters* 14(4):16-21.

Raymond, R. 1987. Lecture in Philadelphia, PA. April 29.

R. E. Wright Associates Incorporated. 1986. Company publication,
Middletown, PA.

Reynolds, T. D. 1982. *Unit Operations and Processes in Environmental
Engineering* (Monterey, CA: Brooks/Cole Engineering Division).

Rittman, B. E., and V. L. Snoeyink. 1984. "Achieving Biologically Stable
Drinking Water." *J. Am. Water Works Assoc.* 76(10):106-114.

Roberts, P. V., and J. A. Levy. 1985. "Energy Requirements for Air
Stripping Trihalomethanes." *J. Am. Water Works Assoc.* 87(4):
138-145.

Schilling, R. D. 1985. "Air Stripping Provides Fast Solution for Pol-
luted Well Water." *Poll. Eng.* 17(2):25-27.

Schmidt, S. K., and M. Alexander. 1985. "Effects of Dissolved Or-
ganic Carbon and Second Substrates on the Biodegradation of
Organic Compounds at Low Concentrations." *Appl. Environ.
Microbiol.* 49(4):822-827.

Singer, M. E., and W. R. Finnerty. 1984. "Microbial Metabolism of
Straight-Chain and Branched Alkanes." In *Petroleum Microbiol-
ogy*, R. M. Atlas, Ed. (New York: MacMillan Publishing Co.,
Inc.), pp. 1-60.

Singley, J. E., L. J. Bilello, and M. A. Mangone. 1981. "Design of Packed Columns for Removal of Trace Organic Compounds from Drinking Water." Proceedings of the 23rd Annual Public Water Supply Engineer's Conference, April 21, 1981. University of Illinois at Urbana-Champaign.

Singley, J. E., A. L. Ervin, M. A. Mangone, J. M. Allan, and H. H. Land. 1980. "Trace Organics Removal by Air Stripping." Prepared for the AWWA Research Foundation, Denver, Colorado.

Snoeyink, V. L. 1981. "Adsorption as a Treatment Process for Organic Contaminant Removal from Groundwater." AWWA Seminar Proceedings, Denver, Colorado, pp. 67-82.

Snoeyink, V. L. 1983. "Control Strategy—Adsorption Techniques." In *Occurrence and Removal of Volatile Organic Chemicals from Drinking Water* (Denver, CO: AWWA Research Foundation), pp. 155-203.

Solmar. 1974. Data sheet.

Speital, G. E., and F. A. DiGiano. 1987. "The Bioregeneration of GAC Used to Treat Micropollutants." *J. Am. Water Works Assoc.* 79(1):64-73.

Sullivan, K., and F. Lenzo. 1983. "Strip Those Organics Away." *Water Management* December, pp. 46-47.

Sutton, P. M. 1986. "Innovative Engineered Systems for Biological Treatment of Contaminated Surface and Groundwater." In Proceedings of the 7th National Conference on Uncontrolled Hazardous Wastes (Superfund 1986), Washington, DC.

Symons, J. M., J. K. Carswell, J. Demarco, and O. T. Love. 1979. "Removal of Organic Contaminants from Drinking Water Using Techniques Other Than Granular Activated Carbon Alone—A Progress Report." U.S. EPA, Drinking Water Research Division, Cincinnati, OH.

Tabak, H. H., S. A. Quave, C. I. Nashni, and E. F. Barth. 1981. "Biodegradability Studies with Organic Priority Pollutant Compounds." *J. Water Pollut. Control Fed.* 53(10):1503-1518.

Treybal, R. E. 1980. *Mass Transfer Operations*, 3rd ed. (New York: McGraw-Hill Book Company).

van der Kooij, I. D. 1983. "Biological Processes in Carbon Filters." In *Activated Carbon in Drinking Water Technology* (Denver, CO: AWWA Research Foundation).

Van Lier, W. C. 1983. "The Kinetics of Carbon Filtration." In *Activated Carbon in Drinking Water Technology* (Denver, CO: AWWA Research Foundation).

Verschueren, K. 1977. *Handbook of Environmental Data on Organic Chemicals* (New York: Van Nostrand Reinhold Company).

Wallman, H., and M. D. Cummins. 1986. "Design Scale-up Suitability for Air Stripping Columns." U.S. EPA, Office of Research and Development, Washington, DC.

Werner, M. D. 1985. "Predicting Carbon Adsorption of Organics from Humid Air." Proceedings of the ASCE Environmental Engineering Specialty Conference (New York: American Society of Civil Engineers).

Wilson, J. T., L. E. Leach, M. Henson, and J. N. Jones. 1986. "In-situ Biorestoration as a Ground Water Remediation Technique." *Ground Water Mon. Rev.* 6(4):56-64.

Worthly, J. A., and R. H. Moser. 1984. "Removing Solvents to Restore Drinking Water at Darien, Connecticut." In *Experiences with Groundwater Contamination.* AWWA Seminar Proceedings, Denver, CO.

Yaniga, P. M. 1982. "Alternatives in Decontamination for Hydrocarbon Aquifers." *Ground Water Mon. Rev.* 2:4.

Yaniga, P. M., and W. Smith. 1985. "Aquifer Restoration: In-situ Treatment and Removal of Organic and Inorganic Compounds." In *Ground Water Contamination and Reclamation* (Bethesda, MD: American Water Resources Association), pp. 149-165.

Yaniga, P. M., and W. Smith. 1986. "Aquifer Restoration Via Accelerated In-situ Biodegradation of Organic Contaminants." In Proceedings of the 7th National Conference on Uncontrolled Hazardous Wastes (Superfund 1986), Washington, DC, pp. 333-338.

Yehaskel, A. 1978. *Activated Carbon Manufacture and Regeneration* (Park Ridge, NJ: Noyes Data Corporation).

Zanitsch, R. H. 1979. "Control of Volatile Organic Compounds Using Granular Activated Carbon." Presented at Air Pollution Control Association, 10th Annual Meeting, September 26-28, 1979. Gatlinburg, Tennessee.

INDEX